Split and Splice

Split and Splice

A Phenomenology of Experimentation

HANS-JÖRG RHEINBERGER

The University of Chicago Press
Chicago and London

The University of Chicago Press, Chicago 60637
The University of Chicago Press, Ltd., London
© 2023 by The University of Chicago
Published 2023
Printed and bound by CPI Group (UK) Ltd, Croydon, CR0 4YY

32 31 30 29 28 27 26 25 24 23   1 2 3 4 5

ISBN-13: 978-0-226-82530-4 (cloth)
ISBN-13: 978-0-226-82532-8 (paper)
ISBN-13: 978-0-226-82531-1 (e-book)
DOI: https://doi.org/10.7208/chicago/9780226825311.001.0001

Originally published as *Spalt und Fuge: Eine Phänomenologie des Experiments* © Suhrkamp, 2021. English version by Hans-Jörg Rheinberger.

Library of Congress Cataloging-in-Publication Data

Names: Rheinberger, Hans-Jörg, author.
Title: Split and splice : a phenomenology of experimentation / Hans-Jörg Rheinberger.
Description: Chicago : The University of Chicago Press, 2023. | Includes bibliographical references and index.
Identifiers: LCCN 2022032888 | ISBN 9780226825304 (cloth) | ISBN 9780226825328 (paperback) | ISBN 9780226825311 (e-book)
Subjects: LCSH: Science—Experiments. | Science—Philosophy. | Knowledge, Theory of.
Classification: LCC Q182.3.R54 2023 | DDC 507.2/4—dc23/eng20221013
LC record available at https://lccn.loc.gov/2022032888
♾ This paper meets the requirements of ANSI/NISO z39.48-1992 (Permanence of Paper).

Contents

Figures

Introduction

The present building blocks of a *Phenomenology of Experimentation* deal with the scientific experiment in its different manifestations. The title is chosen with caution: the aim is to delineate the *shapes* and contours that scientific experimentation has acquired historically and that have come to dominate the field of the empirical sciences since the nineteenth century. Such an endeavor can be seen as an integral part of a historical epistemology at large.[1] The aim is not to present a philosophical determination of the essence—whatever that may be—of scientific experimentation. Nor am I concerned with an attempt at presenting phenomenology as an experimental philosophy.[2]

What is at stake is a consistent assessment of experimentation as a knowledge-generating procedure. The different facets of the shapes of the experiment shall be dealt with from an infrascopic and from a suprascopic perspective so as to render visible what is usually overlooked with respect to experimentation, either because it remains below the threshold of perception or because it lies beyond it. To do this, I take experimental systems as a starting point. They are in themselves already complex units of modern experimentation, which I have described extensively elsewhere.[3] Micrological aspects of experimentation that could only be hinted at in *Toward a History of Epistemic Things* are examined more closely in part I of the book: the production of traces, the construction of models, the ways of making things visible, the forms of grafting, and note-taking. They will be juxtaposed and characterized in their peculiarities and interrelations. The second part of the book will focus on the relations that experimental systems develop among them. The characteristic temporal, spatial, and narrative dimensions of these articulations will be traced. The concepts that serve as guidelines in the investigation will be presented by way of examples, and they will be developed and tested for their usefulness on materials taken from the history of twentieth-century life sciences.

The first part of the book thus deals with aspects of the experimental infrastructure and its particular materialities. Chapters 1 and 2 consider the constitution of epistemic objects, or epistemic things, as they take shape in the course of experimentation.[4] They are the necessary correlates of nascent scientific concepts for an epistemology anchored in scientific practice. The first chapter takes a close look at traces, the volatile products of the encounter between the means and the targets of investigation. Generated at their intersection, these traces testify to what happens at the center of an experiment. Here, we are moving in the core region of epistemic things, as far as they can be made accessible in an experiment. The discussion will then turn to data, the consolidated forms of these traces, and the transformations and transpositions that characterize this transition. Only by this transition is a space created, the data space, in which the experimentally generated traces can be manipulated and processed. This space will be the focus of chapter 2. It is a space in which the objects of epistemic interest acquire shape through data condensation and become temporarily reified. One of the central and best known of these shapes is traditionally dealt with under the concept of model. Widely differing variants of models play a decisive role in all experimental sciences. They stand in the center of the second chapter. The third chapter deals with the graphematic procedures through which data are configured and models represented accordingly. They act as visualizations of epistemic configurations. However, epistemic things are also, and necessarily, embedded in technical environments. Experimental systems derive their dynamics from their reciprocal interplay. Therefore, chapter 4 turns from the objects of investigation to the technologies of inquiry and exploration. It attempts to explain how research technologies become integrated in the course of experimentation in such a way that they promote the phenomenotechnical effects described in the first two chapters. The manipulations on the part of the technical environment of the experiment are discussed under the notion of grafting. Finally, the fifth chapter of part I addresses the literal activities that make the laboratory a writing space of a special character.

Altogether, from an epistemic as well as a technological perspective, the first part of the book is devoted to the circumstances and conditions that give the process of experimentation its structural cachet and make it a device from which novelty can emerge. That is, ultimately, what lies at the heart of a research experiment: it is oriented toward the generation of new knowledge. Two further aspects of a phenomenology of experimentation that belong in this context have already been discussed extensively in the fourth part of my previous book on the history and epistemology of the experiment, *An Epistemology of the Concrete*: the phenomena of preparation, and the interface between the apparatus and the object.[5] Consequently, I shall not devote separate chapters to them, but will refer to these issues wherever appropriate, particularly in chapters 2, 3, and 4.

If the first part of the book deals with aspects of experimentation that concern the constitution and the dynamics of epistemic things and the technical conditions under which knowledge objects take shape and lead to experimentally induced knowledge effects, the second part focuses on aspects that place experimentation in larger spatial and temporal contexts, and consequently presents experimentation in its expansive dimensions. The transition from the micro-level to the meso-level of experimentation, the passage from an isolated to a multi-meshed, reticulated, regional, and chronotopic way of looking at things, is the consistent follow-up step for a phenomenology of experimental practice that aspires to present things from a perspective "from below." This brings into focus the temporal and spatial relations of experimental systems with each other, their synchronicity and diachronicity, as well as their embedding in historical conjunctures.

Chapter 6 deals with the shapes of time in which the dynamics of experimentation are realized, thereby simultaneously producing these shapes. I will contrast them with the shapes of time that are characteristic of artistic activities. The figurations of time beyond chronology that are connected with experimental forms of knowledge acquisition—the lives of their own—have so far had a rather shadowy existence in epistemology. As is briefly mentioned in chapter 4, experimental systems can form ensembles. We can apostrophize them as cultures of experimentation oriented toward the creation of encounters that are relevant for the generation of knowledge. Experimental systems form families that can dominate a knowledge field for a shorter or longer period. Chapter 7 traces the fate of one such experimental culture—the in vitro culture of biochemistry that shaped the course of the life sciences in the twentieth century. At the same time, the chapter explores the concept of culture in its material dimensions and its relevance for the organization of fields of knowledge. The first part of chapter 8 centers on the narrative dimension that appertains to scientific experimentation and its progressions. The second part of that chapter inquires into the narrative character of a history of science that originates from experimental practice and must therefore lay claim to an experimental dimension of its own activity. Chapter 9 returns once more to the core of the scientific process of knowledge generation and shows that it conceals a certain unruliness, without which the boosts of novelty embodied in the experiment would not occur and would not be able to unfold. In the last instance, we have to do with the wild moment in scientific making and thinking. The tenth chapter, finally, is a short eulogy of the fragmentary character inhering in all experimentation and dominating, in different configurations, the practice of the natural as well as the historical sciences. This completes the arc back to the first part of the book. The work concludes with a short postscript on rhythmanalysis instead of a summary.

In the remainder the introduction, a few words on historically situating such an approach to experimentation in the context of twentieth-century phenomenology may be in order. Ernst Cassirer's *Phenomenology of Knowledge*,[6] conceived and published as the third part of his *Philosophy of Symbolic Forms* in 1929, essentially still targeted the mental, *symbolic* forms, in and through which knowledge acquisition in the sciences proceeded in its development over the centuries. In the fourth and last volume of his tetralogy on *The Problem of Knowledge: Philosophy, Science, and History since Hegel*,[7] however, Cassirer took a decisive step further. Here we read: "No matter whether we are concerned with the ideal or the real, the mathematical or the empirical, with nonsensuous or sensuous objects, the first question is always not what these are in their absolute nature or essence, but by what medium they are conveyed to us; through what instrumentality of knowledge the knowing of them is made possible and achieved."[8]

This is the point to which the present description can connect. What needs to be worked out is an appropriate attention regarding the forms of *practice* with its instrumentalities, in which these knowledge processes develop and through which they are realized. In other words, what moves in the foreground is the material constitution of the most prominent procedure that scientific knowledge generation began to make use of in the early modern period, which has since acquired a dynamic that dominates the empirical sciences throughout. In the eighteenth century, speculative schemes, usually called "systems," still governed natural philosophy and natural history, interspersed with occasional confirmatory experiments.[9] Since the nineteenth century, the situation has been reversed: theories and concepts have to accommodate themselves to experimental systems if they claim scientific relevance. The long history of this reversal and its historical dynamic will not be discussed here. The forms of development of experimentation in the course of the centuries, which appear to be at a turning point again today—a turn characterized by the production and processing of data in mass formats—will have to await further study.[10]

I do not inquire into the foundations of the experiment, in the sense of an ontological quest, nor do I aim to go beyond the experiment, in the sense of a metalogical grounding. Rather, I remain with what Gaston Bachelard once aptly called the "scientific real"—*le réel scientifique*.[11] What I seek to expound in all its detail is the particular form of life and way of being of that scientific real we know as the experiment. Like scientific work itself, the experiment's epistemological reflection revolves around visualization—not of something that may lie beneath or beyond, but of articulations in the plane. In this respect I think it is justified to talk of a phenomenology. It is a phenomenology that no longer obeys a logic of expression, according to which the phenomena are always only something secondary, derived; instead it follows a logic of articulation according to which the phenomena gain their sense and meaning through their references to each other. The phenomenological is taken at face value here. The task is not to seek hidden depths behind the appearances. Everything depends on what is in between, not what is beneath. In an early text, Michel Foucault formulated this succinctly: "Original forms of thought are their own introduction: their history is the only form of exegesis they tolerate, and their fate the only form of criticism."[12]

A phenomenology thus understood can neatly affiliate itself with what Gaston Bachelard called the "phenomenotechnique,"[13] or "phenomenography,"[14] of modern knowledge acquisition, and which he understood first as an "extension" of,[15] then as an alternative to traditional phenomenology. In interwar France, Bachelard founded another tradition, that of a knowledge phenomenology, which proceeds not from the intentional manifestations of conscience, but from the forms of appearance of epistemic things as they arise from "experiments on matter" and the "*field of obstacles*" they engender.[16] In "Phenomenology and Materiality," the introduction to his last book on epistemology, *Rational Materialism*, Bachelard characterized classical phenomenology, as it had taken root in France in the footsteps of Edmund Husserl's *Cartesian Meditations*,[17] with the following words: "Classical phenomenology expresses itself complacently in terms of *purposes*. Consciousness is then associated with an entirely directed intentionality. Consequently, an excessive centrality is attributed to it. Consciousness is the center from where the lines of research get dispersed."[18] What he pleaded for instead was a phenomenology that proceeded from the "resistance of matter" encountered in experimentation.[19]

The central theme of the present book is the *experimental situation* in all of its complexity, and not an eidetic reduction, or intuition of essences. It is therefore neither a phenomenology of states of consciousness, nor one of intentional acts in isolation, but rather a phenomenology of a process of conceptualization inextricably intertwined with its target objects, which are the driving force for its development. We are concerned with an experience that manifests itself as experimental practice, in the double meaning—of experience and experiment—of the French *expérience*. In the words of Jean-Toussaint Desanti, what has to be shown is that "every philosophy of consciousness demands, in the last instance, to be superseded by a philosophy of practice."[20] In the present book, however, I am less concerned with a philosophy of practice in general than, much more specifically, with the experimental practice of the modern sciences.

The questions raised here have occupied me persistently over the past two decades, as continuations of the reflections in *Toward a History of Epistemic Things* and *An Epistemology of the Concrete*.[21] A series of preparatory thoughts have been published since then, many of them in obscure places. I revisit them here, develop them further, and connect them under an overarching frame. I hesitate, however, to speak of a systematic connection. "Phenomenology is an elastic method of description," Hans Blumenberg once remarked succinctly; and, as he formulated this with respect to a "phenomenology of history," he saw its task as the "exposure of the arsenal of forms in which history accomplishes itself that leads to an understanding of their structures independently of the contingencies of their facts, albeit *only through them.*" In other words, it is about "taking into and keeping in culture that very field of an 'endless effort,' which can never be completed by the ideal of a deduction."[22] What is envisaged here is therefore not so much a deductive ideal in the sense of specifying the place of the experiment in an idealized picture of the sciences; it is rather the attempt, with respect to experimentation, at a "synthesis and synopsis," an "overview and conspectus" in the sense of the late works of Cassirer on the structure and dynamics of the modern sciences,[23] an effort that remains as a task when it is no longer a matter of deducing the whole from first principles.

To conclude, let me say a word on the title. What does *Split and Splice* have to do with a phenomenology of the experiment? I hope that this will become clear and plastic in the course of the book, although, on good grounds, neither assignment will be used in a terminological fashion throughout the text. That much, however, in advance: The two words stand in for two procedures that are inherent to all experimentation. Splitting represents the movement of breaking up and its more or less irregular results. Deliberately, I do not use the word "analysis," which points to clear divides and straight results. They are seldom realized in an experiment. As a rule, dissecting a piece of nature does not result in clean divisions. Splitting, in contrast, addresses a zone of hunching that can guide an exploratory movement around the nooks and corners of the matter at hand. The complementary movement of splicing abuts parts that do not neatly match. With equal deliberation, I do not use the word "synthesis" here, for the latter points to junctions that let the parts disappear. The joint, in contrast, is a visible trait that marks a linkage, along which the compound assemblage can also be dissolved again. Splitting and splicing thus stand in for the tentative, as the emblems for what we call a trial.

PART I

Infra-Experimentality

Traces

The realm of the infra-experimental spreads out under the hands of the ex-perimenter. In this veritable underworld, if I may call it such, the task is to capture and make manifest the moments, the chains of events, in which epis-temic things gain significance and scientific meaning becomes reified step by step. At stake is the micro-dynamics of the research process. Gaston Bach-elard coined the notion of "micro-epistemology"[1] to describe his own view *from below*. Neither the traditional epistemological conception of induction nor that of deduction will be of help to us here. Likewise, we have to relin-quish Charles Sanders Peirce's idea of abduction, although it already comes closer to what is at stake.[2] I will speak of subduction, and this in a very pe-culiar sense: here, we are moving in a realm that is framed in such a way that novelty can come about inadvertently, that the unprecedented can be made to happen. It is precisely the realm that lies in the space *between* the agents of knowledge and the objects of their interest. With the advent of the modern sciences, this domain has expanded into an enormous machinery that tends to devour the poles between which it extends and to engender its own world: the underworld of research technologies.

What is actually happening here? How can we grasp what is occurring and what all this extravagance is for? It is certainly not too far-fetched and exaggerated to claim that the original gesture of the modern sciences in the first place consists in trying to make the invisible visible. The whole effort is designed to reveal structures and processes to our eyes or, more generally, to make them accessible to our senses, to make manifest those things that are not disclosed by unmediated observation and thus are not immediately evident. Manifestations of this kind necessarily depend on widely different forms of intervention and manipulation—that is, in the last instance on an instrumentally mediated disturbance. The philosopher and art historian Edgar Wind once expressed this insight as follows: "This intrusion, of which every investigator must be guilty if he wishes to make any sort of contact with his material and to test the rules of his procedure, is a thoroughly real event. A set of instruments is being inserted, and the given constellation is thereby disturbed."[3] The inescapability of such disturbances is the underlying reason for why, in modern scientific knowledge production, such a close connection has developed between science and technology—that is, between the production of knowledge and the technical means on which its procedures rest.

The sciences as we know them today only exist in and through this media-technological landscape. It is, however, not always already present; it has emerged from the process of knowledge generation itself and its extensions.[4] In this sense, we can talk of a technological disposition of the sciences. A techno-epistemic momentum inscribes itself into the modern generation of knowledge.[5] It manifests itself in historically widely variegated forms, most of which, however, turn out to be transient. Think of the inscription devices that revolutionized nineteenth-century physiology and then became completely obsolete in the course of the twentieth century.[6]

That should not, however, tempt us to amalgamate science and technology indiscriminately and to talk about techno-sciences without further ado, as frequently occurs in the literature of studies in science and technology.[7] I shall show that it will be rewarding to distinguish between the epistemic and the technical moments of the sciences instead of mingling them, and to study both in their interactions. At the same time it will become clear how the technical and the epistemic assume—and promote—each other reciprocally and how, through precisely that interaction, they confirm each other in their distinctness.

Traces

In my earlier book, *Experiment, Differenz, Schrift*,[8] I began to use the concept of the trace to characterize the primary products of this interaction. I mobilized it as a prerequisite for an adequate apprehension of another concept, that of representation and the imagery that attends it. In doing so, I referred to Jacques Derrida's grammatological reflections.[9] Another useful reference in this context can be found in the deliberations of the art historian Georges Didi-Huberman.[10] It seemed rewarding to probe the concept of the trace in the context of characterizing the process of experimentation. I have not been alone in attempting this.[11] By "traces" we mean a form of material manifestation—a form of palpability—that is characterized by the following peculiarities. First, a trace is more elementary and more rudimentary than what we usually understand as a representation. Second—to borrow from Peirce's semiotic vocabulary—its nature is indexical:[12] it is the primary manifestation of an epistemic thing. Third, it predates the distinction between writing and imaging, which are our traditional forms of representation in the wake of and following the trace, the raw material of the experimental semiosis. In preceding that distinction it exposes the "asemic kernel" of that semiosis,[13] which survives only as a residual category in writing and as an image.

There is a conundrum inscribed into the structure of the trace. As trace, it is the trace of something, but that something is always absent. As trace, its character is that of a substitute. With that, it refers to an episteme that basically operates on the terrain of displacement and deferral. Derrida pinpoints this character exactly when he says: "But a meditation upon the trace should undoubtedly teach us that there is no origin, that is to say simple origin; that the questions of origin carry with them a metaphysics of presence."[14] Where we have to do with scientific research, we can still radicalize this figure. This something, the supposed origin of the trace, is absent not only in the sense of no longer being here, but in a much stronger sense: it ever was before. We cannot catch the thing that generates the trace *in flagrante*. Were this possible, we could save ourselves the whole experimental effort. The epistemic take on the world rests on this constitutive belatedness. Of course that also means we have to proceed on the assumption that recursivity is inscribed into the structure of the systems of empirical investigation. In accordance with their whole construction, they are "recurrent," to borrow a notion that Bachelard has used to describe the *New Scientific Spirit*.[15] On the other hand, that implies a "transgression." The young Jean Cavaillès expressed this figure in his review of a lecture of Léon Brunschvicg at the Second University Conferences in Davos in 1929 as follows: "In every moment of science, the reason that is orientated toward the real—not that kind of reason which satisfies and reassures the logicians—transgresses itself."[16] We encounter here a figure of temporality with roots in the unfinishable and the a-teleological. The process of knowledge generation rests not only on the inherent necessity of positing ever new suppositions, but in being compelled to discount them again and again in one and the same movement.

From this perspective, it should be possible to come a good step closer to understanding the fundamental cultural technique of experimentation. And that, in turn, should open up ways to determine more specifically what goes on, from an epistemic viewpoint, in the material transformation processes of scientific experimentation without always already moving in the auspices of image and writing. That at least was, and is, the idea of approaching an understanding of the dynamics of scientific knowledge generation from the perspective of experimental systems. This is done not least in the hope that the pursuit of experimental systems, which themselves can be seen as traces of the history of the modern sciences, will also communicate to the historiography of science the experimental dynamics that it is seeking to identify in the sciences. Edgar Wind has already pointed out that the historical sciences and the natural sciences are faced with a comparable situation: "namely, that every discovery regarding the objects of their inquiry reacts on the construction of their implements; just as every alteration of the implements makes possible new discovery."[17] I shall return to this point again in part II of the present book.

If one chooses such a starting point, then the key task is to reflect on the epistemic and technical constitution of such trace-generating experimental systems and the experimental environments or landscapes that they form. But it is exactly that reflection on the experimental mediation—in other words, the knowledge apparatus in all its complexity and intricacy—which inserts itself between the knowledge-seeking subjects and the objects of knowledge; it is exactly the new world of traces that it engenders in these middle grounds—the variegated and expansive *eigenspace* of the trial—that failed to find its proper place in the perspective of classical, traditional epistemology and philosophy of science. Strangely enough, it has not found its place in the self-perception of the sciences, either. If scientists speak as "spontaneous philosophers,"[18] the instrumental mediation as a rule completely disappears from view; what counts alone is the result, the finding. Consequently a conceptual effort toward leverage of a phenomenology of experiment is required that ascribes a central role to the tools of this activity and makes them available for reflection on matters of science.

The present book is devoted to this task. Its starting point is the occurrences and events in the experiment. The perspective that underlies it privileges neither the knowing "I" nor accomplished knowledge; rather, it positions itself in between. All experimentation disembogues in generating traces. This does not usually happen in a completely haphazard or fortuitous fashion, but moves on pathways, in trajectories. Bachelard's writings contain an early example of a sentence that neatly highlights this systemic character: "The time of the adaptable patchwork hypothesis is over, and so is the time of fixation on isolated experimental curiosities. Henceforth, hypothesis is synthesis."[19] All experimentation moves along two different epistemic axes, depending on whether it is about the exploration of spatial structures or the determination of temporal sequences. Experimenting with structures revolves around procedures of compression on the one hand and dilatation on the other: what is too big to be perceived has to be miniaturized, what is too small has to be magnified. And the experimental exploration of processes revolves around procedures of acceleration and of retardation: what is too quick to be perceived is slowed down, and what is too slow is accelerated. These procedures represent instructions for the generation of traces. At the same time, they supply options as to how these volatile traces can be transformed into durable data, and for ordering and condensing them into data clusters in such a way that patterns can emerge that give contours to the investigated phenomenon. Robert Merton has pointed to the importance of "establishing the phenomenon," as he called it.[20] I will return to these procedures in more detail in chapter 3. The transformations that they entail are, however, neither monotonic nor of a generic nature, but are variegated in character and depend on the specific technologies through and in which they are realized. These technologies will be dealt with in chapter 4.

Radioactive Traces

At this point I would like to concretize what I have said so far and briefly present a research technology that decidedly contributed to the molecularization of biological research around the middle of the twentieth century: the technology of radioactive labeling. It spread through all of the life sciences between 1950 and 1980 and played a crucial role in the sequence analysis of nucleic acids. One of its designations is radioactive tracing, which refers directly to what characterizes it: the production of traces.

At the time of William Crookes in England, Henri Becquerel in France, and Hans Geitel and Erich Regener in Germany at the very beginning of the twentieth century, physicists began to occupy themselves with the following phenomenon: radioactive substances, which had moved by then into the center of attention in the natural sciences and beyond, elicited light flashes, so-called scintillations, when they hit appropriate screens.[21] Before long it was realized that these light traces could be used as a measure for the radiation intensity of the respective substances. The Hungarian physical chemist Georg von Hevesy called them "indicators" or "tracer elements." He was one of the first to employ them to follow the metabolism of substances within the organism.[22] The concept of "indicator" that von Hevesy used in his works from the 1920s refers to the circumstance that in these experiments, a radioactive substance "indicated" in two respects—transitively and intransitively. On the one hand, the radioactive decay traces indicated the path the substance took through the body. One could literally map that passage. On the other hand, a minuscule radioactive admixture served as an index for the complete stream of that substance. A mere "trace" of this substance thus sufficed to indicate its sum total, rendering a signal without causing radiation damage in the respective tissue. In the relevant literature, the notion of "signal" is often used more or less synonymously with that of "trace." The notion of signal reminds us unmistakably of the information technology context in which the development of radioactive counting was embedded.[23] In terms of a further differentiation that generally remains implicit, however, we can say that the notion of signal is mainly used if the event—the radioactive decay—is considered from the perspective of the substance from which the signal is emitted. With the notion of "trace," however, one has an eye on the effect left by the indicator.

Let us now look at a more recent example from the late 1970s in more detail. Figure 1.1 shows one of the first published photograms of a radioactively labeled nucleic acid sequence gel. It is a figure from the paper in which Frederick Sanger and his colleagues Steve Nicklen and Alan Coulson in Cambridge first published their new sequencing method.[24] It shows part of the DNA sequence of the bacteriophage named PhiX174, one of the smallest viruses then known to infest bacteria. The barcode-like representation is the result of a procedure that can be addressed not only as a system of the generation of traces, but also as a system for their registration. The notion of chromatography that is used in technical terminology refers to two elements: to a graphism of sorts, and to color. As on a writing surface, spots (i.e., bars) are arranged in a linear fashion. As can be seen in the figure, they are used to help visualize the linear sequence of building blocks of a nucleic acid as a ladder of stripes following each other at more or less regular intervals. Sanger and his colleagues developed the procedure at the Laboratory for Molecular Biology (LMB) in Cambridge in the second half of the 1970s. With its technological potential it marked the beginning of genome research—that is, the sequencing of nucleic acids on a large scale. Its institutionalized form emerged with the Human Genome Project less than a decade later.[25]

A short description of the technique, avoiding technical terms as far as possible, could read as follows: In a first step, one half of a DNA double strand is re-synthesized in a complementary fashion using an isolated and purified enzyme that usually synthesizes nucleic acids in the cell. In contrast to the synthetic process in the living cell, however, a chain termination of the newly synthesized strand is effected statistically after each addition of a new building block. To achieve that requires the preparation of four reaction mixtures in parallel, each containing, in addition to the four regular building blocks (comprising the bases adenine, thymine, cytosine, and guanine: A, T, C, G), a small amount of a radioactively labeled modified building block that, if

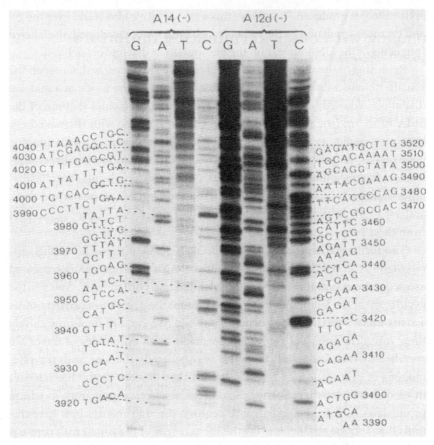

FIGURE 1.1. Sequence gel with a partial sequence of the DNA of bacteriophage PhiX174
Source: Frederick Sanger, Steve Nicklen, and Alan R. Coulson, "DNA Sequencing with Chain-Terminating Inhibitors," *Proceedings of the National Academy of Sciences of the United States of America* 74 (1977): 5464, fig. 1.

incorporated, prevents further synthesis. In a second step, the four reaction mixtures containing the DNA fragments that are end-labeled with one each of the radioactive base-analogues are run on a porous gel consisting of a network of polyacrylamide, whereby the fragments become sorted according to their size. This happens by applying an electric voltage to the gel in which the DNA fragments—chemically seen poly-anions of varying charges—travel over different distances in the gel according to their sizes. This achieved, the resulting bands are visualized. One can look at the gel under UV light that is being absorbed by nucleic acids. In this way, the DNA "writes" or traces its own sequence. For further evaluation, an autoradiogram of the plate (as it is aptly called) is produced. To achieve this a ray-sensitive film is laid over the gel and becomes irradiated by the radioactive bands. Once developed, the labeled fragments of the DNA appear on the photogram as discrete black bands.

Returning to the distinctions made above, we are concerned here with the visualization of a molecular structure that operates in the mode of a massive dilatation: although the DNA fragments represent molecules that are of the order of 10^{-9} m (the nanometer realm), we can now see with the naked eye their mass-agglomerations as discrete bands. In a way, the chromatographic plate orchestrates the technology itself in a pictorial fashion: it realizes a molecular magnification. It presents itself as an ingenious molecular graphic that relies exclusively on the physical and chemical properties of the nucleic acids, and yet yields information about a molecular order that cannot be derived from these properties: a sequence.

Two further remarks may be appropriate here. On the one hand, this representation can rightly be regarded as an analytic one. The DNA is presented as a sequence of building blocks: it is separated into components. And yet, the peculiarity of the technique lies in the fact that the analysis is carried out by a statistically interrupted molecular synthesis. We can see here how closely these categories interlock. On the other hand, such ensembles of traces are specified by their indexical character, as is generally the case with the phenomena of preparation—I have dealt with their history in the life sciences in more detail elsewhere.[26] They range from anatomical preparations whose prime dates back to the seventeenth century, through microscopic preparations that became emblematic for the life sciences of the nineteenth century, to biochemical preparations whose development dominated the twentieth century.

To start with, the relevant parts of preparations materially coincide with components of the sample. A chromatogram such as that described above fixates the epistemic object in its own materiality and in a discrete form, as the material reconfiguration of a molecule that has become legible in this manner—in other words, as a second-order epistemic thing. This kind of material continuity distinguishes preparations from models, which I will discuss more extensively in chapter 2. In contrast to preparations, models presuppose a change of medium and unfold their potential in the semiotic space of the iconic and the symbolic, to remain with Peirce's differentiation.

At this point, it remains to emphasize that these graphematic configurations, through which the technology of radioactively labeling molecular structures and their functions in the living cell operates, are by no means an esoteric peripheral matter in a corner of molecular biology. This micro-technology has spanned a tracing space, which has dominated the molecular visualization of the phenomena of the living for several decades.[27] The Atomic Age, the Molecular Age, and, as we shall shortly discuss, the Age of Information genuinely began resonating with each other through this technology. The biochemical and biophysical experimental systems of molecular biology in general, and of molecular genetics in particular, would have been not only unrealizable, but also unthinkable, without tracer technology. It opened an experimental space of possibilities that transcended the options of classical analytical chemistry by several orders of magnitude, from the micro- to the nano- and even the pico-range, and consequently moved the realm of the molecular into the domain of the visible. In conjunction with a series of other advanced analytical technologies—such as ultracentrifugation, electron microscopy, and X-ray structure analysis—it created a horizon in which an arsenal of traces could be produced that made the life sciences appear in a completely new light. I will discuss the combination of such technologies and their impact in more detail in chapter 4.

Data

Sequence gels, like other experimental traces, are transitory, evanescent things. Moreover, one does not see anything on such a gel to start with, and consequently it cannot be worked on as such. An additional manipulation is necessary to make the bands visible. The usual procedure—as described above—is autoradiography. Whereas the preceding gel electrophoresis still involved the material components of the probe and distributed the molecules of interest along a spatial axis—like a sequence that is supposed to be one-dimensional—the autoradiogram rests exclusively on the rays that the probe emits, and transposes the otherwise invisible intensity pattern onto a sensitive film. This renders it visible and at the same time preserves it in a durable manner. Beyond that, the radiogram can be worked on: it can be annotated, for example. These are all properties that we expect of data in common discourse. In other words, we are concerned here with a transformation of volatile traces into sustainable data. Data are therefore not at all original; unlike what the name suggests, nothing is "given"—it is all the result of a process. Traces are fixed in a suitable medium and can be processed further only in this durable manner.

Phenomenologically, scientific practice manifests itself in widely different forms. The gesture of transposition appears, however, to be common to all of them. In principle, it says that things get removed from a given context—of their embedding or usage—and transposed into a context that allows us to observe or to handle them for the purposes of knowledge generation. Here

we could review the different forms of scientific practice—observation, classification, quantification, isolation, purification, analysis, synthesis, to name
just some of them—and determine more precisely the distinct forms such a
transposition can assume. In the present book I concentrate on the forms it
assumes in the experiment.

To remain with our example: by now we are a good way off from the virus
that infests a bacterial cell; from the isolation of the DNA that is supposed to
be its hereditary material; from the separation of its double strand; from the
synthesis of complementary nucleic acid strands of different length on one of
its original strands by an enzyme, a polymerase; and from their subsequent
segregation in the field of a gel electrophoresis. But even then, this is not
the end of the chain of transformations. It continues, now as a transposition,
into the data space that was entered with the autoradiogram. It already had a
strange status: it was the icon of something that remained invisible until the
image emerged and could be addressed as the image of a banding pattern that
had remained hidden in the gel.

Now, the four lanes can additionally be annotated, merged, and transposed into a chain of four letters (fig. 1.2). They stand for the two purine (A
and G) and the two pyrimidine (C and T) bases whose succession determines
the sequence, and along with it the genetically relevant informational character of the nucleic acid. The illustration shows the complete sequence of
bacteriophage PhiX174 with a length of somewhat more than 5,000 nucleotides. It was the first completely identified base sequence of a phage-genome
that had to be composed of a whole series of sequence gels, on each of which
only fragments of a restricted length could be represented. This composition
alone shows that the symbolic data space thus entered definitely transcends
the boundaries of a single autoradiogram and a single experiment.

The sequence has now landed in the space of writing, and thus in the
space of the possibility of discrete processing. All materiality now appears
stripped away. The asemic kernel of the trace, which was already superimposed on the autoradiogram by a one-dimensional order, is now pushed even
further into the background. But something has been added as well: the sequence becomes legible as a series of triplets that, in the language of molecular biology, contains the information for the fabrication of a cellular product,
a protein molecule. The sequence acquires biological meaning.

With the transformation of traces into data that, in the end, are usually
coded as arrays of digits, letters, or pixels, a decisive transition is being effected. One can address it also as a transition from the graphematic space
of experimentation to the representational space of data processing.[28] Possibly the most important consequence of this transition is that the trace-like

```
GAGTTTTATCGCTTCCATGACGCAGAAGTTAACACTTTCGGATATTTCTGATGAGTCGAA
AAATTATCTTGATAAAGCAGGAATTACTACTGCTTGTTTACGAATTAAATCGAAGTGGAC
TGCTGGCGGAAAATGAGAAAATTCGACCTATCCTTGCGCAGCTCGAGAAGCTCTTACTTT
GCGACCTTTCGCCATCAACTAACGATTCTGTCAAAAACTGACGCGTTGGATGAGGAGAAG
TGGCTTAATATGCTTGGCACGTTCGTCAAGGACTGGTTTAGATATGAGTCACATTTTGTT
CATGGTAGAGATTCTCTTGTTGACATTTTAAAAGAGCGTGGATTACTATCTGAGTCCGAT
GCTGTTCAACCACTAATAGGTAAGAAATCATGAGTCAAGTTACTGAACAATCCGTACGTT
TCCAGACCGCTTTGGCCTCTATTAAGCTCATTCAGGCTTCTGCCGTTTTGGATTTAACCG
AAGATGATTTCGATTTTCTGACGAGTAACAAAGTTTGGATTGCTACTGACCGCTCTCGTG
CTCGTCGCTGCGTTGAGGCTTGCGTTTATGGTACGCTGGACTTTGTGGGATACCCTCGCT
TTCCTGCTCCTGTTGAGTTTATTGCTGCCGTCATTGCTTATTATGTTCATCCCGTCAACA
TTCAAACGGCCTGTCTCATCATGGAAGGCGCTGAATTTACGGAAAACATTATTAATGGCG
TCGAGCGTCCGGTTAAAGCCGCTGAATTGTTCGCGTTTACCTTGCGTGTACGCGCAGGAA
ACACTGACGTTCTTACTGACGCAGAAGAAAACGTGCGTCAAAAATTACGTGCGGAAGGAG
TGATGTAATGTCTAAAGGTAAAAAACGTTCTGGCGCTCGCCCTGGTCGTCCGCAGCCGTT
GCGAGGTACTAAAGGCAAGCGTAAAGGCGCTCGTCTTTGGTATGTAGGTGGTCAACAATT
TTAATTGCAGGGGCTTCGGCCCCTTACTTGAGGATAAATTATGTCTAATATTCAAACTGG
CGCCGAGCGTATGCCGCATGACCTTTCCCATCTTGGCTTCCTTGCTGGTCAGATTGGTCG
TCTTATTACCATTTCAACTACTCCGGTTATCGCTGGCGACTCCTTCGAGATGGACGCCGT
TGGCGCTCTCCGTCTTTCTCCATTGCGTCGTGGCCTTGCTATTGACTCTACTGTAGACAT
TTTTACTTTTTATGTCCCTCATCGTCACGTTTATGGTGAACAGTGGATTAAGTTCATGAA
GGATGGTGTTAATGCCACTCCTCTCCCGACTGTTAACACTACTGGTTATATTGACCATGC
CGCTTTTCTTGGCACGATTAACCCTGATACCAATAAAATCCCTAAGCATTTGTTTCAGGG
TTATTTGAATATCTATAACAACTATTTTAAAGCGCCGTGGATGCCTGACCGTACCGAGGC
TAACCCTAATGAGCTTAATCAAGATGATGCTCGTTATGGTTTCCGTTGCTGCCATCTCAA
AAACATTTGGACTGCTCCGCTTCCTCCTGAGACTGAGCTTTCTCGCCAAATGACGACTTC
TACCACATCTATTGACATTATGGGTCTGCAAGCTGCTTATGCTAATTTGCATACTGACCA
AGAACGTGATTACTTCATGCAGCGTTACCATGATGTTATTTCTTCATTTGGAGGTAAAAC
CTCTTATGACGCTGACAACCGTCCTTTACTTGTCATGCGCTCTAATCTCTGGGCATCTGG
CTATGATGTTGATGGAACTGACCAAACGTCGTTAGGCCAGTTTTCTGGTCGTGTTCAACA
GACCTATAAACATTCTGTGCCGCGTTTCTTTGTTCCTGAGCATGGCACTATGTTTACTCT
TGCGCTTGTTCGTTTTCCGCCTACTGCGACTAAAGAGATTCAGTACCTTAACGCTAAAGG
TGCTTTGACTTATACCGATATTGCTGGCGACCCTGTTTTGTATGGCAACTTGCCGCCGCG
TGAAATTTCTATGAAGGATGTTTTCCGTTCTGGTGATTCGTCTAAGAAGTTTAAGATTGC
TGAGGGTCAGTGGTATCGTTATGCGCCTTCGTATGTTTCTCCTGCTTATCACCTTCTTGA
AGGCTTCCCATTCATTCAGGAACCGCCTTCTGGTGATTTGCAAGAACGCGTACTTATTCG
CCACCATGATTATGACCAGTGTTTCCAGTCCGTTCAGTTGTTGCAGTGGAATAGTCAGGT
TAAATTTAATGTGACCGTTTATCGCAATCTGCCGACCACTCGCGATTCAATCATGACTTC
GTGATAAAAGATTGAGTGTGAGGTTATAACGCCGAAGCGGTAAAAATTTTAATTTTTGCC
GCTGAGGGGTTGACCAAGCGAAGCGCGGTAGGTTTTCTGCTTAGGAGTTTAATCATGTTT
CAGACTTTTATTTCTCGCCATAATTCAAACTTTTTTTCTGATAAGCTGGTTCTCACTTCT
GTTACTCCAGCTTCTTCGGCACCTGTTTTACAGACACCTAAAGCTACATCGTCAACGTTA
TATTTTGATAGTTTGACGGTTAATGCTGGTAATGGTGGTTTTCTTCATTGCATTCAGATG
GATACATCTGTCAACGCCGCTAATCAGGTTGTTTCTGTTGGTGCTGATATTGCTTTTGAT
GCCGACCCTAAATTTTTTGCCTGTTTGGTTCGCTTTGAGTCTTCTTCGGTTCCGACTACC
CTCCCGACTGCCTATGATGTTTATCCTTTGAATGGTCGCCATGATGGTGGTTATTATACC
GTCAAGGACTGTGTGACTATTGACGTCCTTCCCCGTACGCCGGGCAATAACGTTTATGTT
GGTTTCATGGTTTGGTCTAACTTTACCGCTACTAAATGCCGCGGATTGGTTTCGCTGAAT
CAGGTTATTAAAGAGATTATTTGTCTCCAGCCACTTAAGTGAGGTGATTTATGTTTGGTG
CTATTGCTGGCGGTATTGCTTCTGCTCTTGCTGGTGGCGCCATGTCTAAATTGTTTGGAG
GCGGTCAAAAAGCCGCCTCCGGTGGCATTCAAGGTGATGTGCTTGCTACCGATAACAATA
CTGTAGGCATGGGTGATGCTGGTATTAAATCTGCCATTCAAGGCTCTAATGTTCCTAACC
CTGATGAGGCCGCCCCTAGTTTTGTTTCTGGTGCTATGGCTAAAGCTGGTAAAGGACTTC
TTGAAGGTACGTTGCAGGCTGGCACTTCTGCCGTTTCTGATAAGTTGCTTGATTTGGTTG
GACTTGGTGGCAAGTCTGCCGCTGATAAAGGAAAGGATACTCGTGATTATCTTGCTGCTG
CATTTCCTGAGCTTAATGCTTGGGAGCGTGCTGGTGCTGATGCTTCCTCTGCTGGTATGG
TTGACGCCGGATTTGAGAATCAAAAAGAGCTTACTAAAATGCAACTGGACAATCAGAAAG
AGATTGCCGAGATGCAAAATGAGACTCAAAAAGGAGATTGGCATTCAGTCGGCGACTT
CACGCCAGAATACGAAAGACCAGGTATATGCACAAAATGAGATGCTTGCTTATCAACAGA
AGGAGTCTACTGCTCGCGTTGCGTCTATTATGGAAAACACCAATCTTTCCAAGCAACAGC
AGGTTTCCGAGATTATGCGCCAAATGCTTACTCAAGCTCAAACGGCTGGTCAGTATTTTA
CCAATGACCAAATCAAAGAAATGACTCGCAAGGTTAGTGCTGAGGTTGACTTAGTTCATC
AGCAAACGCAGAATCAGCGGTATGGCTCTTCTCATATTGGCGCTACTGCAAAGGATATTT
CTAATGTCGTCACTGATGCTGCTTCTGGTGTGGTTGATATTTTTCATGGTATTGATAAAG
CTGTTGCCGATACTTGGAACAATTTCTGGAAAGACGGTAAAGCTGATGGTATTGGCTCTA
ATTTGTCTAGGAAATAACCGTCAGGATTGACACCCTCCCAATTGTATGTTTTCATGCCTC
CAAATCTTGGAGGCTTTTTTATGGTTCGTTCTTATTACCCTTCTGAATGTCACGCTGATT
ATTTTGACTTTGAGCGTATCGAGGCTCTTAAACCTGCTATTGAGGCTTGTGGCATTTCTA
CTCTTTCTCAATCCCCAATGCTTGGCTTCCATAAGCAGATGGATAACCGCATCAAGCTCT
TGGAAGAGATTCTGTCTTTTCGTATGCAGGGCGTTGAGTTCGATAATGGTGATATGTATG
TTGACGGCCATAAGGCTGCTTCTGACGTTCGTGATGAGTTTGTATCTGTTACTGAGAAGT
TAATGGATGAATTGGCACAATGCTACAATGTGCTCCCCCAACTTGATATTAATAACACTA
TAGACCACCGCCCCGAAGGGGACGAAAAATGGTTTTTAGAGAACGAGAAGACGGTTACGC
AGTTTTGCCGCAAGCTGGCTGCTGAACGCCCTCTTAAGGATATTCGCGATGAGTATAATT
ACCCCAAAAAGAAAGGTATTAAGGATGAGTGTTCAAGATTGCTGGAGGCCTCCACTATGA
AATCGCGTAGAGGCTTTGCTATTCAGCGTTTGATGAATGCAATGCGACAGGCTCATGCTG
ATGGTTGGTTTATCGTTTTTGACACTCTCACGTTGGCTGACGACCGATTAGAGGCGTTTT
ATGATAATCCCAATGCTTTGCGTGACTATTTTCGTGATATTGGTCGTATGGTTCTTGCTG
CCGAGGGTCGCAAGGCTAATGATTCACACGCCGACTGCTATCAGTATTTTTGTGTGCCTG
AGTATGGTACAGCTAATGGCCGTCTTCATTTACATGATGGTCGACTTTATGGCGACATTC
CTACAGGTAGCGTTGACCCTAATTTTGGTCGTCGGGTACGCAATCGCCGCCAGTTAAATA
GCTTGCAAAATACGTGGCCTTATGGTTACAGTATGCCCATCGCAGTTCGCTACACGCAGG
ACGCTTTTTCACGTTCTGGTTGGTTGTGGCCTGTTGATGCTAAAGGTGAGCCGCTTAAAG
CTACCAGTTATATGGCTGTTGGTTTCTATGTGGCTAAATACGTTAACAAAAAGTCAGATA
TGGACCTTGCTGCTAAAGGTCTAGGAGCTAAAGAATGGAACAACTCACTAAAAACCAAGC
TGTCGCTACTTCCCAAGAAGCTGTTCAGAATCAGAATGAGCCGCAACTTCGGGATGAAAA
TGCTCACAATGACAAATCTGTCTCACGGAGTGCTTAATCCAACTTACCAAGCTGGGTTACG
ACGCGACGCCGTTCAACCAGATATTGAAGCAGAACGCAAAAAGAGAGATGAGATTGAGGC
TGGGAAAAGTTACTGTAGCCGACGTTTTGGCGGCGCAACGTGATTGGGCTATCCAACCTGCA
AATTTATGCGCGCTTCGATAAAAATGATTGGCGTATCCAACCTGCA
```

FIGURE 1.2. Representation of the total DNA sequence of bacteriophage PhiX174

Source: Bruce Alberts et al., *Molecular Biology of the Cell* (New York and London: Garland, 1983), 104, fig. 3-8.

results of an experiment are brought into a form in which they can be stored and consequently retrieved and processed, compared with other data, and compressed into patterns. There are good reasons to believe that the durability thus achieved is the most important feature of the transposition from traces into data. Such a fixation is a prerequisite for their new mobility in the data space, a character the traces in the space of experimentation lack. In this respect Bruno Latour once aptly spoke of "immutable mobiles."[29] However, these entities that can be moved through space and time also carry a potential danger for science. As data, the traces are secured and disambiguated. They are deprived of their precariousness, their tentative nature, their preliminary character, the measure of indeterminacy that is inherent in the traces.

In the context of the genome projects that began to be carried out from the late 1980s onward, the research technology whose principles have been used here to exemplify the concepts of trace and data has become automatized, and its efficiency has concomitantly been greatly increased. One decisive development was the replacement of radioisotopes as trace generators with fluorescent molecules. Fluorescence labeling involving the use of four different colors—one for each of the nucleotide bases—has also relocated the transition from traces to data into the machinery itself, has enabled its automation and thus made the whole procedure invisible, with the actual steps being hidden, if not effaced. In the context of a micro-epistemology such as that presented here, however, it is of vital importance not to ignore these transitions altogether.

A consequence of the resulting exponential accumulation of sequence data was the bioinformatic structuring of the molecular genetic data space. By now the structuring of these repositories has widened to a task that has itself acquired an epistemic character. Today it amounts to a rearrangement of the data space as a secondary space for experimentation. Let us not forget, however, that the data space has exposed this character *in nuce* from the very beginning. The different protocols of experimenting on and with the primary material have already a dual character. On the one hand they serve to fix the results of an experiment and gather them synoptically; on the other they bring with them a tentative ordering and reordering of the data and, consequently, of experiments *in papyro*. We will hear more about this feature of the data space in the following, particularly in chapter 5. We cannot ignore, however, that this dual character has acquired a new dimension in the data space of bioinformatics. To begin with, the main focus was on the arrangement of and access to data stemming from heterogeneous contexts.[30] By now, experimenting *in silico* and the related underlying algorithms have acquired yet another new dimension.[31] The information concept in this scientific field has also shifted in meaning. It was introduced in what is now seen as the classic age of molecular biology to express the fact that the nucleotide sequence of a gene specifies the amino acid sequence of a protein and thus a biological function.[32] Now, in the age of genomics, it is being eclipsed by a concept of information that essentially refers to the plethora of collated data in the structured and formalized space of bioinformatics. Managing the bioinformatics space and relating its elements to each other requires consistent and unified standardizations. It is slightly ironic that the respective protocols are described as "bio-ontologies."[33] Yet they prepare the ground on which experimenting *in silico* can develop as a reality *sui generis*.

In the present context it is not necessary to discuss this information concept in more detail. One thing, however, should be added here. The world of epistemic things, whose knowledge is at stake in all these procedures, still persists behind these spaces populated by experimental traces and data, which these and related technologies such as nucleic acid and protein chips help to span. It is the stereo-chemical property of base complementarity (C pairs with G, A with T) on which all these procedures rest. Their experimental observation, which goes back to the work of Erwin Chargaff in the late 1940s, was one of the landmarks of the rising new molecular biology.[34] It provides a good example for demonstrating how an epistemic thing—in the present case, the structure of deoxyribonucleic acid—can become transformed into a technical object and consequently into a new tool of analysis.[35] If the leading technology of classical genetics rested on the hybridization of whole organisms, the key technologies of molecular genetics rest on the hybridization capacity of nucleic acid molecules. There, as here, the structure of an organic interface is exploited, and the capacity for synthesis is transformed into a tool of analysis. This distinguishes these technologies from other biological techniques whose interfaces run between an apparatus and an organic structure. I will return to the latter in chapter 4.[36]

This introductory chapter has dealt with the experimental generation of traces and the consequences that derive from their transformation into data in the context of processing experiments. Traces and data form the basis of this micro-phenomenological exploration of the experimental process. We have discussed them using the example of radioactive tracer technology, a procedure that simultaneously allowed a glimpse of the ramifications and extensions resulting from the creation of the respective knowledge spaces. The chapters of the first part of this book will pursue these ramifications further from the perspectives of modeling epistemic things, of visualizing experimental results, of grafting new technologies onto experimental systems, and of writing protocols. This will require reference to the concepts of trace and datum over and again.

2

Models

The targets of scientific research—the epistemic things—can take widely, even bewilderingly different forms according to the realm they belong to. Nevertheless, a certain order can be brought into this diversity. In the first chapter I have already drawn a basic distinction between the configuration of epistemic objects in the space of traces—that is, in immediate vicinity to the experimental course of events—and their configuration in the data space: the distinction between graphematic and representational exposition. Preparations that have long played a central role in the biological sciences, in medicine, in materials research, and in geology, mineralogy and chemistry, and which differ widely depending on the respective field, can still be counted as belonging to the space of experimental traces.[1]

Moreover, in almost all of the sciences work with models figures prominently. Models can be assigned to the data space and its reifications. I shall consider them in more detail in this chapter. In doing so, I will concentrate on the relation between models and the experiment.[2] That means that I will only consider models that are involved in an empirical research process, and only in relation to that involvement. Whereas preparations, in one form or other, participate in the materiality of the phenomena being investigated, models presuppose a change of medium. A common feature of all these models is thus that they rely on an ontic cut with respect to the epistemic things of the first order.[3] Of course, this says nothing as yet about the medium itself, whether from a presupposing or an excluding viewpoint. Models can be of a purely formal nature (mathematical or logical), or schematic or diagrammatic, and require only paper and pencil for their representation. They can, however, also draw on other materials that allow us to use them as working models, to handle them in space and time and to tinker with them.[4] In addition, computer models and simulations have become ubiquitous and are now used in laboratories around the world.[5]

The following discussion will focus on the construction of models in the process of experimentation—that is, in the empirical sciences. Research experiments such as those done in this realm are rarely singular events; as a rule, they form series in the course of which epistemic things take shape, as already observed and described by Ludwik Fleck in his fundamental study on the *Genesis and Development of a Scientific Fact*.[6] In their elaborated form, such series of experiments extending over longer periods of time can be addressed as experimental systems.[7] If one asks more precisely what goes on in such systems or trajectories of experiments, a general answer could be that an epistemic target, a phenomenon of interest to an experimenter, is being coupled with a technical surrounding in such a way that their interaction exposes it and leaves traces behind. In chapter 1, I described in detail how using a particular research technology—in this case, radioactive tracing— generates a technophenomenon that becomes manifest in the form of experimental traces. If one aims to work with them, one has to give a durable shape to these primary legacies of the experiment. This can be achieved either materially or in the form of an "inscription."[8] It is this step that requires a change of medium. And it is here, as well, that the concept of data comes into play: traces rendered durable and storable in repositories of different kinds can be addressed as data. This step, in turn, makes the next one possible, the step around which this chapter revolves: connecting data with each other. The result is that models appear as data aggregations or assemblies.

From such a strictly bottom-up perspective, which is simultaneously a perspective of science in the making, models can be addressed—as a first approximation—as data clusters or data textures in data space. Basically, they allow us to focus synoptically on a multiplicity of data, and thus create the illusion of being able to *see* the whole. Such a totalization goes along with a peculiar form of inversion. As the anthropologist Claude Lévi-Strauss once remarked, "with a scale model *the knowledge of the whole precedes that of its parts*. And even if this is an illusion, the reason of the procedure is to create or sustain the illusion. . . ."[9] Lévi-Strauss hits on a decisive point here. Reductive synthesis is the way in which such an inversion of part and whole is achieved. From the perspective of the epistemic objects at which they are directed, models are always reduced models. From the perspective of the traces generated in the experiment, they are synthetic. In the graphematic space of the experiment, one never can get hold of the whole. How far one is able to go with such selective syntheses depends on a permanent process of negotiation that is phenomenon-oriented on the one hand, and historically determined on the other.[10] The Harvard ecologist Richard Levins once formulated this with respect to models in population biology as follows: "The difference between legitimate and illegitimate simplifications depends not only on the reality to be described but also on the state of the science."[11]

The strength of models, and at the same time their weakness, resides in simplifications of this kind—which are, however, already complex in themselves. The weakness of such simplifications consists in their inviting us to forget the illusory moment inherent in them; their strength is that they form a flexible scaffold that can react sensibly as a whole if its constituent data are altered, exchanged, or reconfigured. In other words, they form a system in data space. It is precisely their systemic character that carries with it the possibility that punctual interventions can have consequences for the model as a whole. However, that also means that they open up the possibility of something like trial actions. The questions resulting from such interventions into the model can then be fed back into the continuing stream of the experimental production of traces. This procedure initiates a circuit that implies a permanent change of media in both directions: from the experimental system to the model and vice versa.

I do not claim that this description exhaustively captures forms and ramifications of models in the sciences. But from the perspective of a micro-epistemology adopted here it is first and foremost a matter of such bottom-up models. According to Margaret Morrison, models that take their starting point from theories follow an inverse top-down pathway.[12] Consequently, we are then no longer dealing with models *of*, but models *for*.[13] Nevertheless, such models also undergo the media change described here, at least insofar as they are intended to be embedded in an empirical context at all and not just to remain deliberately speculative. They therefore require a concretization with respect to the epistemic objects with which they are brought into connection. Yet another, different kind of model plays a role in Alfred Tarski's mathematical model theory.[14] These model conceptions, however, are beyond the scope of this book, along with the topic of the usage of models in art and architecture.[15]

Models from below give rise to further differentiations. I will refer to some of them in the following sections, in particular the distinction between functional and structural models in the context of the characterization of biomolecular processes. As in chapter 1, I will develop my reflections using concrete examples. All three modeling media mentioned at the beginning of this chapter—paper, materials of different kinds, and the discretized space of the virtual—have played a role in the experimental work that led to the elucidation of key molecular mechanisms for the life sciences and for molecular genetics in particular: the biosynthesis of proteins, and in particular the structure and function of the cellular organelles by which this synthetic process is realized, the ribosomes. In what follows, I will concentrate on this historical example. The ribosomes, as these organelles have been called since around 1960, are the protein fabrics of the cell—of the a-nuclear, unicellular organisms such as bacteria as well as of the nucleated cells of multicellular organisms. The historical overview given here focuses on the simpler bacterial cell, in particular the model organism *Escherichia coli*; the examples are taken from the original literature in this field.

Model Organisms

First, a preliminary remark about the concept of the model organism.[16] Although model organisms have only been called such since the second half of the twentieth century, they were nevertheless already established as experimental or research organisms in the early twentieth century, and they played a decisive role in the life sciences in general and in genetics in particular over the past hundred years. In this compound form, the concept of model takes on a meaning that is not covered by the model concept discussed in the course of this chapter, and therefore requires a short explication.[17] From an evolutionary perspective, it appears as a peculiar characteristic of living beings that the differences between them are based on historical contingency. These differences can only be determined case by case and not deduced from general principles. Nevertheless, modern biology assumes that fundamental similarities exist between different organisms and that, once originated, they have been retained by the different species in the course of further evolution. This situation confronts biologists with two problems. The first is that it is essentially a matter of inductive generalization to determine how ubiquitously a particular character of living beings presents itself. There are no a priori reasons for biological generalities. The second problem is that in the exploration of these more general features, decisions have to be made: a certain characteristic may be more easily accessed and more readily substantiated in its specific features in one particular organism than in another. It is exactly here that the concept of a model organism comes into play. Model organisms are in certain respects particularly apt—that is, more or less ideally suited objects of investigation. Such suitability consists, first, in that a certain biological phenomenon is to be observed in them in a particularly pronounced fashion; second, and more importantly, that they reveal themselves in this respect as being especially convenient to manage. As a rule, and paradoxically, such ideality has *material* consequences. One intervenes into certain properties of an organism in order to standardize it, as is the case with the breeding of pure lines or the generation of defined gene combinations in genetic model organisms. *Model* organisms are therefore always organisms that have been *modified* for the purposes of research. That determines their character as research tools. They embody previously acquired knowledge. In this sense, model organisms as such are not objects of investigation or epistemic things of which a model is to be constructed. Rather, they belong to the technical conditions of an experimental system. They are research technologies of a biological character. In comparison to other research technologies, we can say that the interface with the epistemic objects for whose investigation the model organisms are kept is internal to them.[18] I will return to the problem of the interface between the targets of research and the research technologies in more detail in chapter 4.

Models of Protein Biosynthesis

Knowledge about the complicated structure of the bacterial ribosome developed by means of an interaction between the establishment of experimental systems and the construction of models of widely different kinds, some of which will be presented as examples in this section.[19] These models circulated as a common good in the discourse of the community of ribosome and protein synthesis researchers, and they were permanently put on trial and re-modeled over decades.[20] Empirical models are not "immutable," but highly changeable "mobiles."[21] The protein synthesis models stand here as prime examples of the process of molecular model building as it evolved in the history of molecular biological research from its beginnings to the present. Molecular biology serves here as an example of an empirically oriented science.

I shall now present the ribosome models in a more or less chronological fashion, as they appeared in the research literature. My account will proceed in three steps: I will look at functional models, then at structural models, and finally at computer models. The latter generally combine structural and functional aspects. We will see that research models such as those considered here are not only representations of epistemic objects that rest on data condensation; as temporary reifications of epistemic things, they also play a temporary role as research devices for generating new knowledge. This means that they are able to oscillate between an epistemic and a technical function. I will return to this point again below. This intermediary role also refers to a further aspect of models: their function as agents or actors.[22]

FUNCTIONAL MODELS

The biochemical and biophysical elucidation of the cellular structures that first entered cell biology as microsomes and were later re-baptized as ribo-

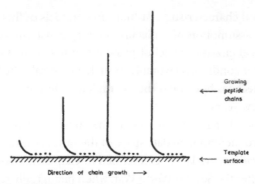

FIGURE 2.1. Simple "model" of protein synthesis as of 1953
Source: Charles E. Dalgliesh, "The Template Theory and the Role of Transpeptidation in Protein Biosynthesis," *Nature* 171 (1953): 1028.

somes because of their high content of ribonucleic acids dates back to the early 1940s. That history began with the identification of a class of small particles in the cytoplasm of mammalian cells and of yeast by the groups around Albert Claude at the Rockefeller Institute in New York and Jean Brachet at the Free University of Brussels.[23] Because of their micro-dimension, the light microscopes available back then no longer showed them as visible structures; their representation was delegated to the differential fractionation of the cell sap in the gravitational field of the ultracentrifuge. Here, they formed the last fraction containing material that could be sedimented. Still smaller molecular complexes and molecules remained in the supernatant. In contrast to the mitochondria, the next bigger class of uniformly centrifugable cell compartments, the microsomes did not show a regular pattern of then measurable enzyme activity. Their metabolic function remained a riddle for the time being. The only thing that was clear was that they mainly consisted of ribonucleic acid and protein.

It was only about a decade later that several groups—among them Paul Zamecnik's at Massachusetts General Hospital in Boston—started following up the suggestion of Brachet and his colleagues that the microsomes obtained from rat liver or yeast might be the cellular sites of protein biosynthesis. One of the earliest and most rudimentary functional models shows—in the utmost abstraction—a simple linear so-called template, a kind of master molecule or matrix, along which the protein threads were supposed to be successively assembled (fig. 2.1). The figure stems from a publication by the British biochemist Charles Dalgliesh in 1953. With the exception of the linearity of the order of synthesis directed by a template, this paper model is not further specified with respect to its molecular character, and it does not contain any further structural characteristics. At that time, models of this kind that were coupled to the assumption of molecular energy consumption began to replace the classical enzyme model of protein synthesis—that of a reversal of proteolysis under conditions favorable to such a reversal. The latter had been dominant for decades but would then completely disappear from the repertoire within a few years.

The data underlying these, as well as the subsequent functional models, again rest on a condensation in data space of the primary traces resulting from the method of radioactive labeling described in chapter 1. Their collection became possible after the Second World War, when biologically relevant radioactive isotopes such as carbon, hydrogen, sulfur, and phosphorus became available for biological and medical investigations. This meant that biomolecules such as amino acids and nucleotides could now be labeled. With suitable filtering and precipitation methods, the products of their reactions could be isolated from the reaction mixtures and counted, first with sensitive Geiger counters, and later with scintillation counters based on photo-multiplication. Here they became transformed into counts—that is, into numbers, the data on which all these models relied. These data are of a numerical nature, in contrast to the data derived from autoradiography, as described in chapter 1. What was needed to gather these data were kinetic experiments carried out using different isotopes and their combinations. On this basis the functional model of protein biosynthesis could be differentiated step by step.

It would, however, still take as much as a decade before a series of molecular components, most of them unsuspected, could be identified and assumed to play an important role in the process of condensation of amino acids into polypeptides.

Two ribonucleic acids that became known around 1960 under the names transfer RNA and messenger RNA played a particularly prominent role in this respect. The unstable messenger RNA was identified as the template for the assembly of the building blocks of proteins. Stable transfer RNAs, in turn, were to accomplish the concatenation process according to a code composed of three nucleotides specifying the respective amino acids. It became increasingly clear that the ribosomes presented a complex and highly asymmetric assemblage. Around this time, researchers throughout the rapidly growing number of laboratories active in the investigation of protein synthesis had managed to consistently separate ribosomes of both prokaroytic (bacterial) and eukaryotic origin into two subunits, a small one and a large one. The two units could be reassembled so that they were able to incorporate radioactive amino acids into proteins in the test tube. After extended in vitro experiments, combined with

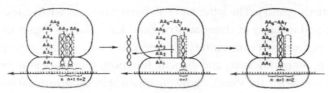

FIGURE 2.2. Functional model of protein biosynthesis from 1964
Source: James D. Watson, "The Synthesis of Proteins Upon Ribosomes," *Bulletin de la Société de Chimie Biologique* 46 (1964): 1419, fig. 20.

the application of the antibiotic puromycin that had been found to interfere with the process of protein biosynthesis, two different binding sites for transfer RNA could be functionally distinguished. Countless experiments of this kind, allowing for the representation of partial reactions of the process of peptide bond formation, led to a still purely formal but much more detailed functional model of the bacterial machinery of protein synthesis. The functional model shown in figure 2.2 represents the state of the art around the mid-1960s.

James Watson, who researched and taught at Harvard University at that time, published this cutting-edge synopsis of the functional sequence of the steps of protein synthesis as of 1964. The model assumes protein synthesis to be a cyclic process during which one amino acid after another is added to a growing peptide chain, called a polypeptide. The model also assumes that the site of peptide bond formation is on the large subunit of the ribosome. The precise succession of amino acids that are being bonded is determined by the sequence of triplets—the codons—of the messenger RNA, the latter being associated to the small subunit of the organelle. The process of synthesis as a whole is mediated through transfer RNA molecules. They recognize the co-dons with one of their extremes (their anticodon), and on the other extreme they present the corresponding amino acids to the enzymatic center in such a way that an energy-dependent condensation reaction can take place.

The publication of Watson's first textbook of molecular biology—*Molecular Biology of the Gene*—made graphical, schematized functional models of this kind into a hallmark for the visualization of living processes at the molecular level.[24] These models emphasized an image of the microscopically invisible molecular structures as a form of mechanisms. The experimental apparatus of this new science predominantly produced the type of mechanistic phenomenologies which transform molecular processes that are optically inaccessible into mechanical scenarios. Within a decade, richly illustrated textbooks of this kind replaced the biochemical textbooks of the first half of the twentieth century. The latter had still been based on written text and the language and imagery of chemical structural formulae—a legacy of the second half of the nineteenth century. However, the new models not only had a mediating, communicative, and didactic function; they were also a means of helping to *condense* the steadily growing amount of experimental results and knowledge components related to these complex molecular processes into a synthetic overall picture. It was a picture that could be grasped at first sight, and that suggested further, experimentally accessible questions on the basis of these synoptic premises. One could also say that with respect to the experiments that had suggested them, they exhibited the character of affordances. They functioned not only as overviews but also as instruments of orientation for research in the coming decades.

One could be tempted to see this kind of imagery as a restriction, as a deficiency of such models, and indeed it produces an illusion, in Lévi-Strauss's sense, occasionally also called a "fiction" in the literature.[25] But this restriction—one could even say malapropism—of molecular processes reveals itself at the same time as an advantage. On its basis, with the molecular states thus fixed, expectations can be formulated and then addressed experimentally. This allows the circularity between model and experiment already mentioned to become palpable again, in all its ambiguity. The model serves as an indirect source for an iterative process of the production of new experimental traces that, when transformed into data, can be reexamined for their compatibility with the existing model. Consequently, they get incorporated into the model, refine it, modify it, or occasionally question it in its entirety. What counts in the end is less its representational character, in the sense of a closeness to nature of any kind whatsoever, than its interventional character.[26] In this sense, one can postulate, as Alain Badiou did in his early book, the *Concept of Model*: "Note that it [the model], as a transitory adjuvant, is actually destined for its own dismantling, and that the scientific process, far from fixing it, deconstructs it."[27]

It is characteristic of the functional models discussed here that they transpose molecular processes with their biochemical and biophysical interactions into an imagery with which we are familiar from everyday mechanical devices. That is also reflected by the use of language—there is continuously talk about molecular mechanisms or molecular machines. They are now also increasingly being discussed in philosophy and in history of science.[28] As a consequence, the biochemical aspects of these processes, such as catalytic properties or energetic aspects, move into the background. This becomes particularly evident when looking at a three-dimensional version of such a model (fig. 2.3). Alexander Spirin, a protein chemist of the Soviet Academy of

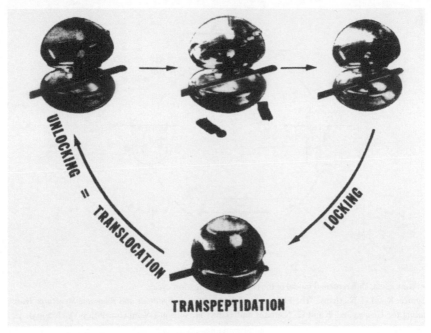

FIGURE 2.3. Metallic functional model of a bacterial ribosome
Source: Alexander S. Spirin, "A Model of the Functioning Ribosome: Locking and Unlocking of the Ribosome Subparticles," *Cold Spring Harbor Symposia on Quantitative Biology* 34 (1969): 201, fig. 4.

Sciences, presented this model at a symposium on protein synthesis in Cold Spring Harbor in 1969.

Between 1960 and 1990, these models and the experimental work mediated through them led to the identification of further molecular factors that play a role in what is called the synthesis cycle of protein fabrication—the repeated loop of peptide bond formation. Through that work, the model became more differentiated, and the cyclical character of the basic molecular operations became more pronounced in these model representations. Moreover, the assumption of an allosteric interaction underlying the whole cycle was included in the model. The representation of the cycle from an actual textbook may help to illustrate this development and to conclude this section (fig. 2.4). Taken together, this series of visualizations gives an impression of how a molecular model, not only embedded in an experimental process but also dragged along with it, can focus research over decades—thereby eventually also neglecting aspects that, under the given experimental conditions, simply do not come into view. Nevertheless, new questions are generated in this way, and the attempted responses continually modify the model, at times merely cosmetically, at times more abruptly, and sometimes to the point of

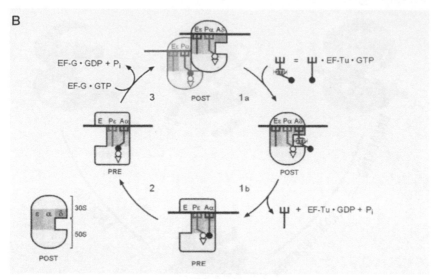

FIGURE 2.4. Differentiated model of the ribosomal elongation cycle
Source: Knud H. Nierhaus, "The Elongation Cycle," in *Protein Synthesis and Ribosome Structure: Translating the Genome*, ed. Knud H. Nierhaus and Daniel N. Wilson (Weinheim: Wiley-VCH, 2004), 327, fig. 8-2B.

its complete abandonment. We need not elaborate further here the abundant details of that history.

STRUCTURAL MODELS

During the 1960s and 1970s, the cellular organelle that lies at the center of the model building process described here was dissected into its nucleic acid and protein components. One component after another was isolated and characterized by sophisticated biophysical and biochemical procedures. On the one hand, efforts were concentrated on determining the amino acid and nucleotide primary sequences and their three-dimensional folding structure. On the other, the challenge was to understand how the dozens of components were connected to each other, how they interacted, and what dependencies between them could be observed if the organelle were taken apart in the test tube and then reassembled from its components. At the same time, efforts were made to determine the contours and surface characteristics of purified particles and their subunits with the help of electron microscopy.

All in all, one can follow the emergence of a complex pattern composed from the traces and the resulting numerical and optical data that were obtained by a whole series of research technologies and that became condensed in these models and combined into a synopsis. The techniques included the isolation of high-molecular components, protein and ribonucleic acid sequencing with the associated determination of the secondary structures of the molecules, selective chemical cross-linking, the assembly of components, the electron microscopic representation of particles, neutron scattering, and X-ray structure analysis. It is impossible to present them all here in detail.

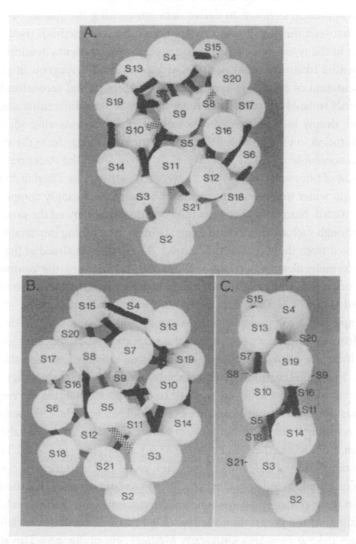

FIGURE 2.5. Protein topography of the small ribosomal subunit of bacteria
Source: Robert R. Traut et al., "Protein Topography of Ribosomal Subunit from *Escherichia coli*," in *Ribosomes*, ed. Masayasu Nomura et al. (New York: Cold Spring Harbor Laboratory Press, 1974), 273, fig. 1.

This complex of experimental arrangements and the resulting multiple and varied data sets resulted in a proliferation of structural models. Figure 2.5 shows an early three-dimensional ball-model of the arrangement of the twenty-one proteins of the small ribosomal subunit Robert Traut and his colleagues assembled at the University of California Davis in the early 1970s. Smaller and bigger polystyrene foam spheres numbered from 1 to 21 represent the differently sized protein components. The connecting rods between them, shaded and hatched in different ways, symbolize the various experimental approaches chosen to derive data concerning the spatial neighborhoods between the components. Among the advanced methods used in this context in the 1970s were the following procedures: covalent chemical cross-linking after treatment of the particles with cross-linking reagents in situ, the representation of neighborhoods as the result of a partial reconstitution of previously isolated components in the test tube, and the differential shielding of more deeply located components from surface reagents after successive reconstitution in vitro—the further a component was away from the surface, the less accessible it proved. After the early 1980s, when the structural determination of ribonucleic acids by methods such as those described in chapter 1 came into wider use, models of this kind became increasingly complex and differentiated. Nucleic acid sequencing enormously enhanced the process.

Although such models exhibit a surface in space, nothing important could be derived from these data in this respect. These models aimed at the internal, intermolecular connections and interactions between the components of the organelle—called quaternary structure by the research community. It resulted from the tertiary structure of the individual components—that is, the shape of the components in three-dimensional space. This, in turn, derived from the secondary structure of the respective macromolecule, and the latter from the primary sequence of the linear thread of each of the molecules. In parallel, various approaches attempted to model the outer shape of the particles. For a while the method of choice for obtaining the required data was almost solely transmission electron microscopy. Here, the scattering traces resulting from the electron beam hitting the particles are visualized on a screen, from where they can be stored as micro-photographic data. The ensuing comparison of a plethora of particles, combined with the superimposition of hundreds of individual images, led to concurring models of their contours in three-dimensional space that differed considerably from each other depending on the preparation method. One of these models was defended by James Lake at New York University Medical School (fig. 2.6). This model, also built from polystyrene foam, showed the two subunits, one above the

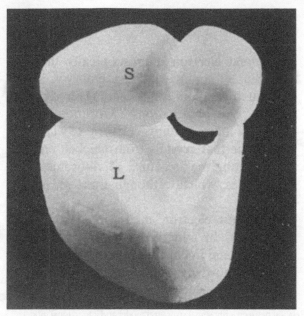

FIGURE 2.6. Structural model of the bacterial ribosome on the basis of electron micrographs
Source: James A. Lake et al., "Ribosome Structure as Studied by Electron Microscopy," in *Ribosomes*,
ed. Masayasu Nomura et al. (New York: Cold Spring Harbor Laboratory Press, 1974), 544, fig. 1.

other, with clearly asymmetric but softened contours—a result of the filtering
process on which the model was based.

    We have now identified the two main parameters underlying the construc-
tion of these structural models: the external shape in three-dimensional space,
and the internal articulation and positioning of dozens of macromolecular
components with respect to each other. Given the total of almost sixty different components, this task could only be approached collectively and over
a longer period of time. In parallel to the analysis of the quaternary structure,
work was also performed on models of the three-dimensional structure of the
individual components, in particular that of the two big RNA strands, each of
which possessed its own complex configuration in space. To build the models,
scholars drew on data about the available primary sequences and the folding
rules and neighborhood analyses of their building blocks because—as with
proteins—particular building blocks that are far apart from each other in the
primary sequence can come near to each other in the spatially folded nucleic
acids. The methods employed were similar to those used for the articulation between the macromolecular protein components. We shall pass over

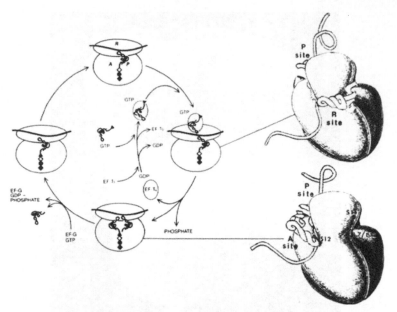

FIGURE 2.7. Relation between a ribosomal structural model and a functional model
Source: Melanie Oakes et al., "Ribosome Structure, Function, and Evolution: Mapping Ribosomal RNA, Proteins, and Functional Sites in Three Dimensions," in *Structure, Function, and Genetics of Ribosomes*, ed. Boyd Hardesty and Gisela Kramer (New York: Springer, 1986), 57, fig. 3.10.

these models here because they do not add significantly to the discussion about models. But this list should clearly illustrate what is involved in such a proliferation of data—concerning tens of thousands of protein and nucleic acid building blocks—to create synopses or data condensations in the form of models.

From the beginning, functional models of the kind I presented in the previous section were related to structural models. The goal was to associate partial functional states of the protein synthesizing ribosome with particular components or regions of the subunits, which are called functional centers, including the corresponding positions of messenger and transfer RNA at the intersection between the two ribosomal halves (fig. 2.7). The example here shows us unmistakably that in an articulated research field in which whole batteries of research technologies, distributed among many working groups, are applied, the participating members of a scientific community begin to communicate through the models involved. This means that at least a part of the communication in such an extensive community no longer follows the track of the primary traces or the data collections and thus the minutiae of every experimental process, but already occurs in a mediated form at the level of synopsis.

As a consequence, there is not only a cycle between models and the experimental production of data, but also feedback *between models* themselves that refer to the same epistemic object but rely on different data sets that illuminate the epistemic thing at hand either at the same angle or from different angles. Such a confrontation of models with each other—usually giving rise to the identification of incongruencies—can again lead to a mutual adaptation of the models in question and to the production of further experimental data. Here we are no longer concerned with the relation of a representation to an alleged phenomenon, where one asks for the meaning of the model and its reference; rather, we are focused on a relation between different representations. In other words, we are dealing with a second-order space in which, to refer to a distinction made by Gottlob Frege, something either makes *sense* or not.[29] Or, to echo Richard Levins's laconic remark, "We attempt to solve the same problem with several alternative models each with different simplifications but with a common biological assumption. Then, if these models, despite their different assumptions, lead to similar results we have what we can call a robust theorem. . . . Hence our truth is the intersection of independent lies."[30]

The function of a model in scientific research is thus not only, and probably not even, predominantly representational: the model is also, and in particular, a tool of further knowledge production. It functions as a temporary reification of the epistemic thing that is the target of investigation. The emphasis here is on *temporary*. For reification always means restriction, abstraction, and simplification as well. Models not only show; they also omit. The epistemic danger of a model consists in making us forget its partial character—which also means its partiality. The multiplication of models is one of the possible epistemic remedies for this restriction. This is exactly the point of Georges Canguilhem's remarks on the fundamental deficiency of models and the danger for associated research when he notes that the function of the model "consists in lending its type of mechanism to a different object, without imposing itself as axiomatic."[31] In their whole mode of being, models therefore embody preliminarity.

COMPUTER-GRAPHIC MODELS

In the late 1970s a report in the research literature on protein biosynthesis aroused attention: ribosomes on the endoplasmic reticulum of the ovary cells of the hibernating South Italian lizard *Lacerta sicula* formed sheet-like crystalline structures.[32] The finding suggested that crystallization of the particles

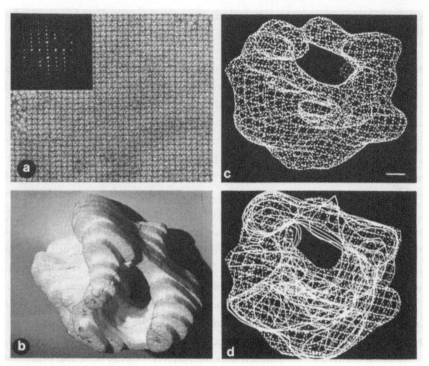

FIGURE 2.8. Crystallography and image reconstruction of ribosomes
Source: Ada Yonath et al., "Crystallography and Image Reconstructions of Ribosomes," in *The Ribosome: Structure, Function, and Evolution*, ed. Walter E. Hill et al. (Washington, DC: ASM Press, 1990), 142, fig. 5.

should also be possible in vitro. Accordingly, starting in the early 1980s, efforts were undertaken to crystallize ribosomes in the test tube and by doing so to clear the way for a structural analysis by X-ray diffraction. The result was a foray into atomic dimensions that had hitherto appeared unthinkable. The early attempts did not look very promising, and widespread opinion in the community of researchers on ribosome and protein synthesis held that it was a wasted effort. They believed there was more to be gained by using such coarser-meshed methods as neutron or small angle scattering. As the existing structural models of the time showed, the particles were highly asymmetric, the number of atoms going well into the hundreds of thousands. The basic question was whether these particles could be brought into a crystalline shape at all, a shape that would lend itself to X-ray structure analysis. At least to begin with, two-dimensional paracrystalline arrays could be obtained and their diffraction pattern could be evaluated and projected onto a physical model by computer-graphic procedures (fig. 2.8).

The image in figure 2.8 comes from a summary report in the year 1990 (that is, a decade after the beginning of the very first crystallization experiments). It would be another decade before three-dimensional crystals of ribosomal subunits could be obtained that were big and regular enough to yield diffraction patterns that led to models with a resolution of around 10 Ångstrom. Meanwhile, two more research groups had joined the worldwide race for an ever higher resolution. Beside Ada Yonath and Heinz Günter Wittmann (Rehovot/Berlin/Hamburg), this involved the group around Venkatraman Ramakrishnan (Salt Lake City) and that around Thomas Steitz (New Haven). The following illustrations from their respective publications show the breadth of options of computer-graphic representations that had been developed over two decades. Figure 2.9 is a stereo representation of the small ribosomal subunit (with a resolution of 5.5 Ångstrom) on which the ribonucleic acid and the proteins are shown in the standard form of their secondary structure—double helix areas for RNA, alpha-helices and beta-sheets for the proteins. Figure 2.10 shows a compact surface representation of the large subunit that rests on structural data with a resolution of 5 to 9 Ångstrom. Abbreviations refer to selected proteins and the binding sites for elongation factors. Finally, figure 2.11 displays an electron density model of the small ribosomal subunit, with a structural resolution of 7 to 12 Ångstrom. At the lower edge of

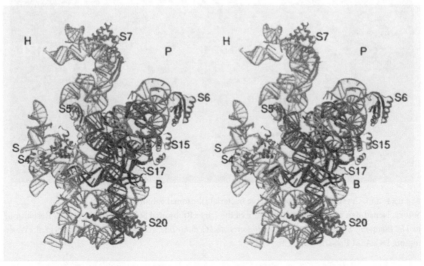

FIGURE 2.9. Crystal structure of a small bacterial ribosomal subunit
Source: Venkatraman Ramakrishnan et al., "Progress Toward the Crystal Structure of a Bacterial 30S Ribosomal Subunit," in The Ribosome: Structure, Function, Antibiotics, and Cellular Interactions, ed. Roger Garrett et al. (Washington, DC: ASM Press, 2000), 7, fig. 6.

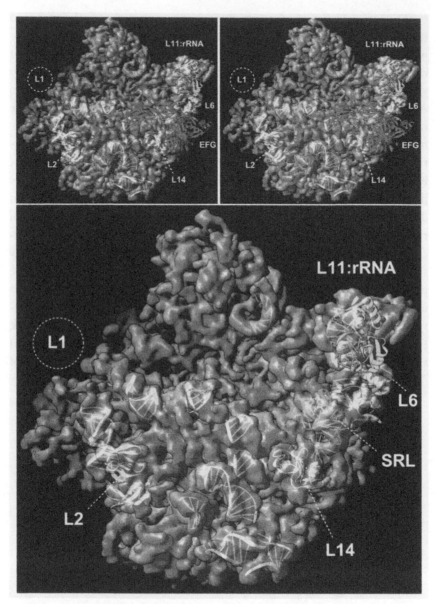

FIGURE 2.10. Crystal structure of the large bacterial ribosomal subunit

Source: Nenad Ban et al., "Crystal Structure of the Large Ribosomal Subunit at 5-Ångstrom Resolution," in *The Ribosome: Structure, Function, Antibiotics, and Cellular Interactions,* ed. Roger Garrett et al. (Washington, DC: ASM Press, 2000), 15, fig. 2.

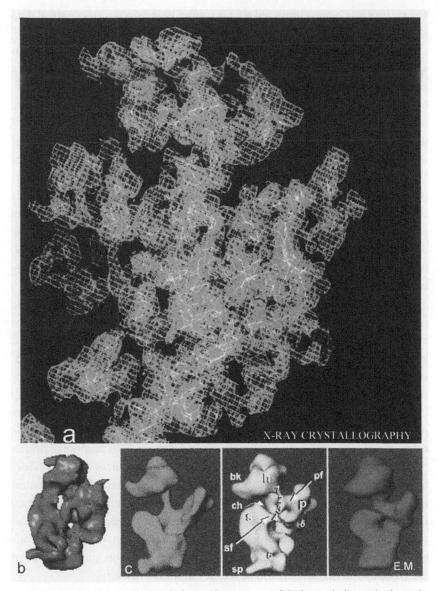

FIGURE 2.11. Identification of selected ribosomal components of the bacterial ribosomal subunits by crystallographic mapping
Source: Anat Bashan et al., "Identification of Selected Ribosomal Compounds in Crystallographic Maps of Procaryotic Ribosomal Subunits at Medium Resolution," in *The Ribosome: Structure, Function, Antibiotics, and Cellular Interactions,* ed. Roger Garrett et al. (Washington, DC: ASM Press, 2000), 23, fig. 1.

the picture it is being compared with a model based on electron optical data stemming from another, independent research technology.

This last image reflects the practice of comparing models resulting from alternative technologies of visualization and therefore resting on different data sets. I have described this procedure briefly already; here again, it shows its potential to drive knowledge acquisition. To create appropriate, trace-generating interfaces between specimen and technology, the different research technologies require different preparation procedures for the probes. All these procedures interfere in one way or another with the configuration of the particles that are labeled native and supposed to exist in their cellular milieu. Native, untouched particles, however, cannot be seen or made visible, which makes the manipulation of their stature unavoidable. The only possibility of gaining a robust assessment of their shape is a permanent triangulation between the different results of such manipulations. In the space of such triangulations, necessarily differing models clash with each other, and the respective representations challenge each other.

As this handful of images illustrates, the resources for computer-graphic modeling are extraordinarily diverse and variegated. The conventions of representation for the secondary structure of nucleic acids and proteins that had been established long before the time of computer modeling became part of its repertoire, like other characteristics of classical three-dimensional molecular modeling such as the space-filling calotte and rod models of chemistry.

One of the most innovative aspects that distinguishes this kind of computer-graphic modeling from the form of a traditional, static model is the free mobility of computer models and their parts in virtual space. This additional option not only offers new possibilities for testing atomic distances and neighborhoods, but also allows representation of functional states and their succession in time. Nevertheless, computer models of this kind cannot be better than the experimentally generated data on which they are based. Ada Yonath, who earned a Nobel Prize for her work on the three-dimensional structure reconstruction of ribosomes in 2009, once replied to my question about the role of the computer in her investigations: "Forget it, it's all chemistry."[33]

SIMULATIONS

To conclude this chapter, let me add a short digression on computer simulation. I mean simulations in the precise sense that computer images are not just based on computationally executed algorithms, but also work with data that are computer-generated. Simulations represent a form of epistemic entities that are qualitatively distinct from the other models considered here.

They belong to the category of models mentioned at the beginning of this chapter and omitted here. As Jean Baudrillard once, and early on, pointedly wrote, simulations are characterized by a "precession" of the model.[34] My concern here has been with models in the empirical sciences that, apart from always being accompanied by a change in the medium, result from experimental traces transformed into data. In contrast, simulations are able to operate on self-generated data, and by doing so, can make us visualize origins and futures that are inaccessible to the real-world experiment—virtual time in virtual space.

Simulations in this sense had their place in the age before the computer, of course, but they had a rather limited radius of action. When computer-implemented, however, they open up an additional space for experimentation in which models themselves can become the central targets of research. Using the exploitation of virtually generated data, they can do this to an extent that transcends by far the work on and with classical models.[35] Whether the technical and epistemic elements configured in this space can be regarded as an alternative kind of experimental system—in silico systems or "simulation systems" or "modeling experiments," as Margaret Morrison and Evelyn Fox Keller call them, respectively[36]—or whether they simply enlarge the technical conditions of wet experimentation is a question that can be posed on the basis of the material discussed here, but not definitely answered. In any case, Franck Varenne is of the opinion that "*integrative software-based computer simulation*" as it developed since the 1970s even surpasses the computer-based implementation of formal models. In doing so, it not only extends the tradition of dealing with formal models as it developed since the beginning of the twentieth century—models *for*, following Fox Keller—with greatly expanded means, thus elevating them to a new level, but also offers the possibility of "experiments with simulation" or "virtual experiments"[37] that even "precede the model as such."[38] This would then be a precession of a second order. It means we would be dealing with experimental systems in virtual space that allow not only for the manipulation of models, but also for their generation and modeling.

We are familiar with the precession of the model in its material and analog form from other areas of culture which are very familiar to us: art and architecture. Here, the relation between the model and the modeled is inverted from the start, if only in forms characteristic for each of these realms. Precisely what the precession of models now happening in the realm of the sciences means for them, in particular for the relation between the sciences and the arts, is beyond the scope of this book and has to remain a question for future discussion.

Making Visible

In this chapter I concern myself with the gesture of making something visible in the context of an experiment. Visualizations were already alluded to in the first two chapters, if only in passing. They shall now become the focus of reflection.[1] Let me stress once more: the visualization of things that are not revealed immediately and are not directly visible to the unmediated eye is the fundamental gesture of the modern sciences in their entirety. The expression of "making visible" directly addresses the moment of making inherent in the procedure. Visualization is always bound to variegated forms of intervening in the target and manipulating its components. It is for this very reason that from the early modern period a tight connection was established between knowledge and the means of its production, between the epistemic and the technical moment of knowledge generation. In this sense—the tool-dependence of scientific knowledge acquisition—we can talk about an inherent technological constitution of natural scientific knowledge production. A technical momentum resides in the innermost core of the epistemic process. As described in the first chapter, it manifests itself initially as a perceivable, recordable trace left by the instrumental intervention. The trace marks the beginning of the process of making something visible.

The trace is a form of manifestation that, as we have seen, lives both from its vicinity to the material being processed and from its proximity to the tools to whose agency it owes its existence. It therefore precedes the critical distinction between image and writing. One must understand its nature if one wants to assess what occurs—and what is at stake—epistemically in the transformation procedures of the experiment. One must also take seriously the technical constitution of knowledge-generating epistemic environments. It is precisely this mediation and, accordingly, the gaze on what goes on in the space *between* the knower and the object of knowledge that has remained epistemologically underexposed, despite the efforts of the contemporary practice turn in sociology, history, and, more recently, philosophy of science.

Interestingly enough, it was none other than Johann Wolfgang von Goethe, who is usually regarded to have emphasized *Schauen* (contemplation) at the expense of intervention, who had already at the turn from the eighteenth to the nineteenth century clearly pointed to these middle and mediating grounds: to the "apparatus" to be handled "under defined conditions" and with the "necessary dexterity."[2] His insights in this respect emerged from his thorough and critical study of Newton's color theory.[3] Unlike most of his contemporaries,[4] Goethe refused to let the mediated quality of knowledge disappear in the result, or to replace it with the apparent transparency of a vista presumed as direct and immediate.

In this chapter I will discuss a number of procedures for making things visible in the laboratory. In contrast to chapter 2, which focused on models as particular forms of reification of epistemic things, the present chapter deals with the actual procedures of visual representation that underlie models as well as preparations, and on which both draw. In a first step, we will deal with the procedures of spatial and temporal compression and dilatation already hinted at in chapter 1. They can be subsumed under the concept of configuration. Second, we shall look at procedures of enforcement or enhancement. And finally, we will discuss procedures of schematization. The chapter is meant as a first outline of a typology of visualization in the sciences.

All the examples of forms of visualization described in the text are taken from the ambit of five experimental techniques that were essential for the emergence of the molecular biosciences around the middle of the twentieth century and that were frequently used in combination. First, there is the use of radioactive isotopes dealt with in both of the previous chapters. The use of these isotopes helped to render visible metabolic processes as well as cell and tissue structures and the macromolecules they consist of. Second, we have the technique of ultracentrifugation, which allows for a representation of different macromolecular components of the cell in isolation from each other. Third, there are the techniques of chromatography, which help to achieve a separation of molecules that are present only in minute amounts, and which therefore fall through the mesh of traditional chemical analytics. Fourth, we have the technique of transmission electron microscopy, with which cellular ultra-structures can be visualized well into the dimension of nanometers. And finally, there is the so-called plaque technique, a molecular genetic procedure with which bacteriophages and their mutants can be made visible. All the examples presented in the following two sections are taken from the original literature. They helped to engender observations and findings that gave new insight into molecular structures and processes—structures and processes that had escaped the traditional technologies of visualization.

Configurations—Compression and Dilatation

Whether we are dealing with the experimental visualization of structures in space or the visualization of processes in time, either compression or dilatation is always in play. It is not an exaggeration to claim that the art of scientific experimentation essentially consists in thinking through and carrying out such contractions and expansions, decelerations and accelerations, thereby generating the necessary means to bring the phenomena at stake into the realm of the visible. Tiny structures have to be blown up, big ones have to be scaled down. Processes that are too quick have to be retarded, those that are too slow have to be sped up.

Let us begin with *spreading*. One of the typical forms of a dilatation of *structures* is optical magnification. Figure 3.1 shows an electron optic magni-

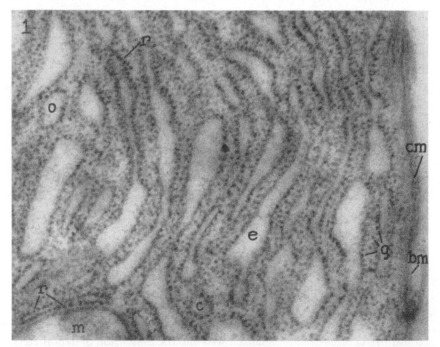

FIGURE 3.1. Electron micrograph of a cytoplasmic region of a rat pancreas cell containing the cell membrane and endoplasmic reticulum
Source: George E. Palade, "A Small Particulate Component of the Cytoplasm," *Journal of Biophysical and Biochemical Cytology* 1 (1955): 59–68, fig. 1.

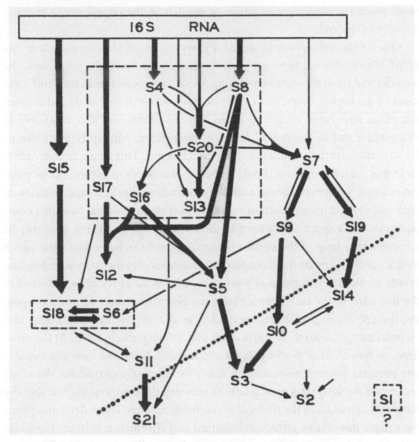

FIGURE 3.2. Assembly map of the small ribosomal subunit of *Escherichia coli*
Source: Masayasu Nomura and William A. Held, "Reconstitution of Ribosomes: Studies of Ribosome Structure, Function and Assembly," in *Ribosomes*, ed. Masayasu Nomura et al. (New York: Cold Spring Harbor Laboratory Press, 1974), 193–223, fig. 1.

fication of a preparation of the cytoplasm of the pancreas cells of a rat. George Palade, who published this cut through a cell with a magnification of 75,000 in 1955, is regarded as the discoverer of those cytoplasmic laminar structures that became known as endoplasmic reticulum. Up to that point this membrane system of eukaryotic cells had remained invisible in the preparations of a histology based solely on light microscopy. That made Palade's publication quite spectacular at the time. We note that the confinements of these cellular structures are studded with small and dark—thus electron-dense—particles that were lined at around this time with the cellular fabrication of proteins: the microsomes. Their visualization as cellular particles was successfully achieved here, also for the first time. In chapter 2 I reviewed their more finely resolved representation by models in the course of the ensuing decades of research.

One of the exemplary forms of a *compression* of *structural data* is the map.[5] We are dealing here not with the structures themselves, but with the data derived from them. When talking about maps, one thinks first and foremost of geological maps, with their occasionally extreme scale reductions. But other structures, including molecular structures, are also amenable to mapping, if not in the typical form of downsizing. Miniaturizing molecular structures further would not make much sense. This example also makes clear that spatial reduction is only a special case of compression. In the present example we are concerned with a condensation through superposition of data sets derived from dozens of experimental attempts—that is, with a compaction of data space. The example shown here (fig. 3.2) is the assembly or reconstitution map of the small ribosomal subunit of bacteria—with which we are familiar already from chapter 2. The map shows the successive binding events through which this assembly proceeds, as far as it can be followed in the test tube. In the early 1970s, Masayasu Nomura succeeded in decomposing the cellular organelle into its nucleic acid and protein components and reconstituting it from those parts into a functional particle again. At the same time, in hundreds of binding experiments he determined how the twenty-one proteins became associated in this procedure. Consequently, the chart represents not only structural relations between the components, but also the temporal dynamics of the molecular assembly process. These determinations were made possible by ultracentrifugation and chromatography for the isolation and purification of the components of the cellular organelle. The gene maps of classical and molecular genetics are other paradigmatic examples of the compaction of structural data.[6]

A typical option for the compression—or dilatation, depending on the case—of *process data* (that is, of a temporal sequence of events) is the curve. In physiology it had already acquired an emblematic character in the nineteenth century. In an exemplary fashion, we can read in the preamble to Emil Du Bois-Reymond's 1848 book on animal electricity: "The dependence of the effect on any circumstance presents itself under the image of a curve, whose x-coordinates represent the magnitude of the arbitrarily altered circumstance, and whose y-coordinates show those of the observed effect."[7]

Whole series of measurements of one or of several variables can be represented synoptically in the course of a curve. In this way, patterns of such courses can be recognized: be it of the distribution or concentration of a substance in space, such as a molecule in a gradient; or the temporal course of a periodically measured parameter, such as in kinetics. If sufficiently complex,

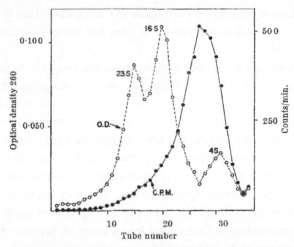

FIGURE 3.3. Sedimentation pattern of unstable RNA of *Escherichia coli* pulse-labeled with radioactive uracil
Source: François Gros et al., "Unstable Ribonucleic Acid Revealed by Pulse Labelling of *E. coli*," *Nature* 190 (1961): 581–85, fig. 8.

such patterns would be sought in vain in the data tables on which the curves are based. In this case, the two-dimensional representation in space is indispensable for pattern recognition.

Figure 3.3 shows a comparably simple-looking, but nevertheless (upon closer inspection) fairly complex example of such a compound curve. The representation results from the interplay between two techniques: radioactive pulse labeling of bacterial cells in vivo, and ultracentrifugation of the cell sap through a sucrose gradient subsequently separated into thirty-five fractions. At the beginning of the 1960s, François Gros and his colleagues at Harvard showed with this experiment that bacterial cells contained an unstable form of ribonucleic acid that had not previously been identified and that subsequently entered the textbooks of molecular biology under the name messenger RNA. We already encountered this molecule in chapter 2 as an integral part of the models of ribosomal function from the early 1960s onward. In the example, the curve with the white circles represents the stable forms of ribonucleic acid in the cell: two ribosomal RNAs (labeled 23S and 16S according to their sedimentation coefficient in the gradient of the cell sap) and transfer RNA (with a sedimentation coefficient of 4S). The superimposed curve with the black dots shows a radioactive peak exactly at the point where the optical density measurement of the ribonucleic acid shows a minimum. It thus refers to the presence of a form of RNA that accrues quickly in the cell at a low concentration and that decays again just as quickly. This is why it can be labeled with a radioactive pulse that does not attain the stable forms of ribonucleic acid.

In an elementary form, the example shows not only a compact arrangement of data in curves, but also the power and knowledge-generating effect of synopsis through the superposition of curves with different but relatable measurement data. This delineates a space of representation in which data can be conjoined into patterns, which in turn can give rise to new experiments and the resulting reconfigurations of these patterns. In this sense, like the models described in chapter 2, these types of representation also function as research instruments. What is being represented here, however, is not a process or a structure within the cell, but the course of an experiment generated in the space of experimentation. Compacted into a curve, it becomes the starting point of new experimental cycles. At the same time, this example reveals that we are not dealing with models here. Not every visualization is a model, and not every model is a visualization. Consequently, models like those described in chapter 2 are not the only configurations that populate the experimental data space. This space is much richer in representational options, many of which are located still nearer to the course of experimentation and can thus themselves become the elements of a model. We can summarize them as pattern formations that rest on the procedures of visualization being discussed in this chapter. These flexible tools can engender representations that do not depict at all in the narrow sense of the expression. Visualization in the sciences is a gesture that encompasses the iconic but is much richer in terms of options.

Enhancement

A further, equally widespread procedure of visualization can be addressed as *enhancement*. The procedures discussed in this section lead, as a rule, to a class of epistemic objects that can be subsumed under the concept of preparation already introduced in chapter 1. Preparations are widespread in the biological sciences, in pharmacology, and in chemistry, and they can assume widely different forms according to the technologies used to create them.[8] Since the seventeenth century, anatomical and physiological preparations based on the scalpel and on conservation procedures belonged to the core of zoological collections of research and teaching devices. With the herbaria, botany developed its own form of dried preparations. With its slides, light microscopy—in particular since the second half of the nineteenth century— led to the creation of a whole universe of durable micro-objects. They served widely different purposes and exerted a variety of functions not only in histology, pathology, embryology, taxonomy, and systematics, but also in geology and the material sciences—thus in a whole panoply of research contexts.[9] Biochemistry and molecular biology of the twentieth century have in turn brought forth a great diversity of analytical preparations, of which the gel-electrophoretic chromatogram shown in figure 3.6 is one example. Characteristic of these latter preparations is that the investigated object undergoes a reconfiguration that would make it completely unsuited to serve as a preparation for microscopic inspection, for example.

To reiterate, preparations are inherently of a variegated nature. Each of these forms of preparation embodies a peculiar representational logic whose specific range and reach has to be sounded out and accomplished historically. What they share as visual forms of objects of knowledge is, first, that because of their indexicality they have to be allocated to the space of traces rather than the space of data. However, they develop in close resonance with the instruments to whose analytic affordances they respond in specific ways and at whose interfaces with the objects of investigation they acquire contours.[10] They can thus be seen as reified intermediates: objects that owe their existence to those spaces between subject and object that bestow their historically variable contours on the realm of the experiment. The knowledge-promoting function of preparations is therefore finite in duration. Just as they can become prominent, so too can they disappear again as soon as more favorable and convenient alternatives are found for them. What is most important in the present case is that their manufacture involves enhancement.

Enhancement means that structures—and sometimes processes as well—are rendered visible by dyeing or by the placing of contrasts, and also occasionally by stiffing (that is, by the strengthening of shapes assumed to be present in the target that are not visible). In the process, the dyes, contrast materials, and means of enforcement become themselves part of that which is to be represented. It follows that deformations are inherent in the procedure. Once again, it is variational multiplication that allows for triangulation and introduces recursivity into the research process.

We have already encountered a form of enhancement that was typical in the molecular biology of the second half of the twentieth century: radioactive labeling. One of the big advantages of working with unstable isotopes is that the radioactive elements do not differ noticeably from their stable variants in their chemical properties and therefore do not alter—given the minimal concentrations in which they are used—the metabolic processes they are deployed to elucidate. Stable atomic components of biomolecules are thus replaced with their unstable isotopes. At the places where they decay, they blacken a ray-sensitive plate placed on the preparation.

Figure 3.4 shows the autoradiogram—sometimes called radioautogram—of an in situ labeled cell of *Tetrahymena*, a free-living unicellular ciliate. The illustration is from a publication by David Prescott, a pioneer in the use of radioisotopes in combination with the micromanipulation of single cells. The cell on the upper part of the figure was exposed for fifteen minutes to a tritium-labeled building block of ribonucleic acid—cytidine, in the event. As can be seen from the picture, the locus of the incorporation of cytidine into cellular structures is concentrated in the nucleus of the cell. On the lower part we see a cell that, after a short pulse label with radioactive cytidine, was allowed to continue its metabolism for another ninety minutes in the presence of unmarked cytidine. Now, the radioactivity incorporated into ribonucleic acid has become distributed over the whole cell. From this it can be concluded that ribonucleic acid is being synthesized in the nucleus and then transported into the cytoplasm. The juxtaposition of the two states manifests the contrast between the site of RNA synthesis (on the DNA in the chromosomes of the nucleus) and the site of its function (on the ribosomes in the cytoplasm). Autoradiograms are a paradigmatic form of enhancement. We have already encountered another of these forms in chapter 1 in connection with the representation of a nucleic acid sequence (fig. 1.1). Today, fluorescence labeling of macromolecules has become a powerful alternative to radio-labeling.[11] Thanks to this new technology, different components of the cell can be differentially labeled, and visualized not only in situ on a fixed preparation, but also in living cells, where dynamic processes can be made visible and followed up by video recording.

Contrast enhancement and dye-fixation of specimens also play a central role in many forms of light and electron microscopy. Histology in the second half of the nineteenth century would have been unthinkable without the procedures developed in this context. Electron microscopic representation of subcellular membranes, organelles and macromolecules in particular, as became possible around the middle of the twentieth century, generally also required contrast enhancement. In electron microscopic specimen preparation, this is often achieved by the addition of heavy metal salts. They attach themselves selectively to the membranes and molecules, thus enhancing their electron diffraction capacity. Figure 3.5 shows the DNA of a burst bacteriophage head in the form of a bundle of loops. The DNA of the bacterial virus is visualized as a single, convoluted thread-like molecule whose length can be measured. The picture, taken by Albrecht Kleinschmidt of the University of Frankfurt, aroused considerable attention on its publication.

In microscopy, contrast enhancement and enforcement of structures are always combined with a corresponding dilatation. The procedure of two-dimensional chromatography we shall now discuss brings molecular structures

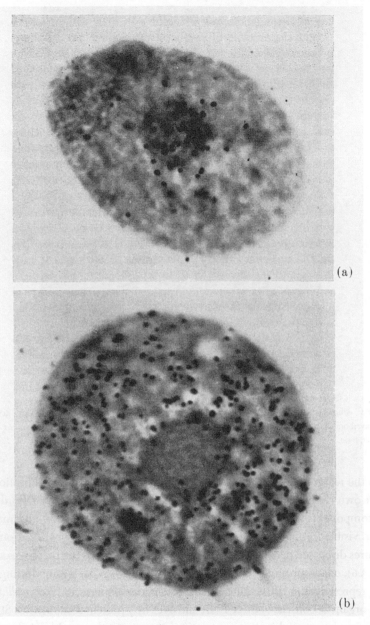

FIGURE 3.4. Autoradiogram of a cell of *Tetrahymena*

Source: David M. Prescott, "Cellular Sites of RNA," *Progress in Nucleic Acid Research and Molecular Biology* 3 (1964): 33–57.

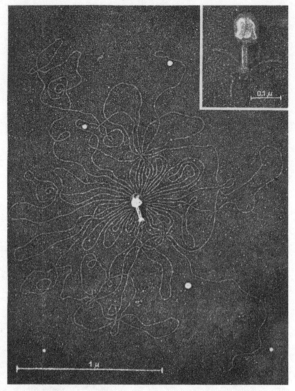

FIGURE 3.5. Electron micrograph of the DNA of Phage T2 at a magnification of x 100,000
Source: Albrecht K. Kleinschmidt et al., "Darstellung und Längenmessung des gesamten
Desoxyribonucleinsäure-Inhaltes von T2-Bakteriophagen," Biochimica et Biophysica Acta 61 (1962):
857–64, fig. 1.

into the realm of the visible and makes them conceivable as magnifications of
their own character. It separates a mixture of macromolecules while ordering
its components according to two parameters—charge and size—and making
them visible through dyeing as distinct spots. As a rule, the visualization procedures discussed here come in different combinations. In the present example
(fig. 3.6), dilatation and enhancement are combined. What we are dealing with
is a representation of the ribosomal protein components of *Escherichia coli*.
The organelle consists of three RNA and over fifty protein components. Strictly
speaking, however, this statement anticipates a result that could only be established through the application of the chromatographic visualization procedure
discussed here.

   With the procedure that was introduced into ribosome research by Heinz-
Günter Wittmann, the protein components of the organelle are separated
first according to their charge in the electric field, and second according to

their molecular weight. However, this alone does not make them visible. The proteins subsequently have to be stained with a suitable dye—methylene blue on this occasion. Otherwise, they would remain hidden and would not appear as dots on the chromatographic plate. They are rearranged according to two physicochemical parameters—size and charge—and reordered in a two-dimensional plane. This rearrangement has nothing to do with their three-dimensional connection in the macromolecular complex of the organelle. Completely different procedures are required to make the latter network visible. One of them has been described earlier in this chapter: the assembly map of the small subunit of the organelle (fig. 3.2). However, with this two-dimensional chromatographic representation, other essential properties such as electric charge distribution, as well as the number and size of the ribosomal

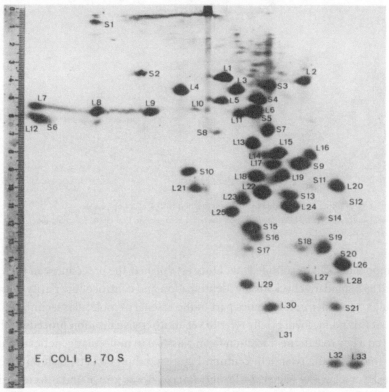

FIGURE 3.6. Two-dimensional electrophoretogram of the ribosomal proteins of *Escherichia coli*. First dimension: 4% acrylamide, basic milieu; second dimension: 18% acrylamide, acidic milieu.
Source: Eberhard Kaltschmidt and Heinz Günter Wittmann, "Ribosomal Proteins. XII. Number of Proteins in Small and Large Ribosomal Subunits of *Escherichia coli* as Determined by Two-Dimensional Gel Electrophoresis," *Proceedings of the National Academy of Sciences of the United States of America* 67 (1970): 1276–82, fig. 4.

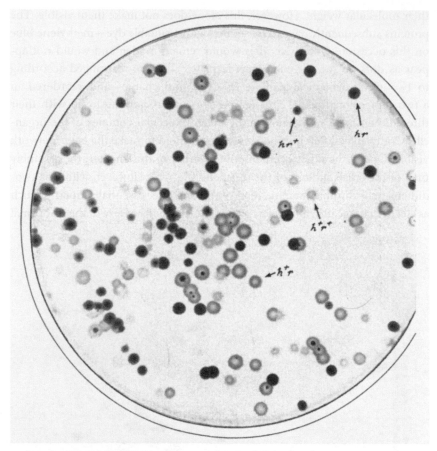

F I G U R E 3.7. Culture of *Escherichia coli* bacteria infected with T2 bacteriophages
Source: Gunther S. Stent, *Molecular Biology of Bacterial Viruses* (San Francisco: Freeman, 1963), 185.

components, can be sounded out. Once established, the procedure can then be used as a standard check for the identification and control of the purity of particular components; it becomes part of the arsenal of molecular techniques.

At this point, I will briefly refer to yet another visualization procedure that played a key role in the transition from classical to molecular genetics around the middle of the twentieth century. Figure 3.7 shows a Petri dish on whose agar layer a lawn of bacteria—*Escherichia coli*—was grown and infected with single bacteriophages. Bacteriophages can be addressed as minute molecular genetic packages that are completely inert on their own. To propagate, they need a bacterial cell. Having infested such a cell, they can induce the bacterium's genetic apparatus to occupy itself with the multiplication of the molecular parasite instead of propagating the bacterium. As a consequence,

the bacterial cell fills with virus particles until it bursts. The phages are freed
and enter neighboring bacterial cells, repeating the cycle.

This principle can be exploited in order to make these minute molecular
packages visible. It works as follows: a Petri dish is prepared with an agar
growth medium on which a regularly distributed bacterial lawn has been
grown. At first there is basically nothing to see on the plate. Then an appropriately diluted solution of virus particles is sprayed over the lawn. The dilu-
tion has to be such that individual bacteriophages, well separated spatially
from each other, fall on the plate. Now the phages can begin their work. They
penetrate one of the adjacent bacteria, multiply, are set free, enter neighboring cells, and gradually form holes in the continuous bacterial lawn. These
holes, which experts call plaques, grow around the point where a single virus
began to multiply. According to the type of phage and its variants, the plaques
can be differently shaded, granulated, stained, and fringed. Their morphology can be distinguished, and the differences can be correlated with the corresponding phage types.

In the present example, *E. coli* bacteria were infected with bacteriophage
T2 and one of its genetic mutants. This was the phage on which the groups
around Max Delbrück and Salvador Luria specialized in the 1940s, thereby
further elaborating the plaque technique. The figure shown here stems from
a textbook of the early 1960s on the molecular biology of bacterial viruses,
written by Gunther Stent, one of Delbrück's later students at the California
Institute of Technology in Pasadena. In the experiment, two types of T2
phages—the double mutant hr and the wild type h^+r^+—were used. Here, "h"
stands for the gene locus "host range," and "r" for "rapid lysis." However, four
different types of plaques can be identified on the Petri dish. The preparation
indicates thus that in the course of virus multiplication, much to the surprise
of the researchers, a genetic recombination between the phage types must
have taken place: in addition to the types hr and h^+r^+, two mixed forms—h^+r
and hr^+—recognizable by their different phenotypes had shown up.

Let us take a closer look at this preparation from the perspective of enhancement. Four aspects can be distinguished. The first concerns the activity
of the biological agents. The plaques that can be seen and inspected by the naked eye are the result of a mass devastation. The myriads of replicated viruses
leave behind innumerable dead bacteria in each of these holes. The second
aspect relates to the very form of visualization. In multiplying themselves and
destroying others, the molecular biological agents produce their own supra-
molecular visibility. Micro-preparations of this kind—that is, particles invisible even under the best light microscopes—give the viruses a macroscopic
shape. The morphology of the plaques, however, has nothing to do with the

molecular morphology of the bacteriophages; rather, it is the result of their molecular genetic properties. As a result of the phenomenological effects of the latter, mutants of the virus can be distinguished from each other.

Third, the form of enhancement to be observed here is accompanied by a dilatation. What can be seen are not single microscopic bacteria or submicroscopic viruses; instead, it is the space occupied by the destroyed bacteria and the multiplied viruses. Fourth, and finally, the preparation teaches us a lesson in molecular genetic research. It not only represents genetic knowledge in a compact form; it becomes itself a tool of research that drives forward the process of molecular genetic knowledge generation. And at the same time it is a research object. It oscillates between technicity and epistemicity. By way of a permanent differential iteration of the experimental procedure, new differences can become visible that, in turn, lead to new characterizations and orient the subsequent course of the experiments. In this case we can see an apparently simple procedure—the plaque technique—developing into a complete experimental system for the investigation of phage genetics. It allows for the visualization of molecular genetic processes in a form which—although, or perhaps because, it has nothing to do with the traditional idea of representation as a depiction—nevertheless renders the molecular processes in question accessible to further investigation.

Schematization

Schematization is a third procedure of visualization that is employed preferentially in the context of a representation of complex processes. It is, therefore, a characteristic of many models. At times, this character prevails to such an extent that modeling and schematization can be used synonymously. Thus, for example, the cycle of protein synthesis shown in figure 2.4 can be addressed either as a model or else as a schema. Nevertheless, the two must be distinguished conceptually. Under the umbrella term "diagrammatics," schemata have recently attracted increasing attention from cultural studies.[12] This is, however, not the place to engage this discussion in more detail. Instead, I will present a few succinct examples of schematization from the recent history of the life sciences.

Like the forms of visualization already discussed, schemata are also polymorphic in appearance. The following examples concern the schematization of biological mechanisms, a genre that has developed an imagery of its own in the course of the twentieth century. This pictorial language developed in parallel to the establishment of molecular biology and has not yet received the historiographic attention it deserves. Its early paradigmatic expression is

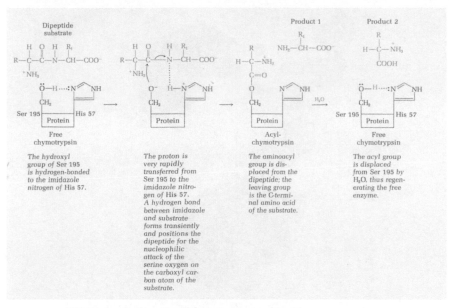

FIGURE 3.8. Mechanism for the action of chymotrypsin on a dipeptide
Source: Albert Lehninger, *Biochemistry*, 2nd edition (1970; New York: Worth Publishers, 1975), 231, fig. 9-15.

to be found in James Watson's *Molecular Biology of the Gene*, from 1965.[13] Well into the 1950s, the formulaic language of biochemistry was the authoritative guideline for emergent molecular biology. In the second half of the twentieth century, new forms of schematization entered not only the textbooks—where they dominate the representation of molecular complexes and of their mode of action as nano-machines today—but also the research literature.

All three examples presented here concern the key biological process of an enzyme reaction. Bio-catalysis is the paradigm of biological processes in metabolism. The transition from the language of chemical formulae to the image regime of schematization is well illustrated in figure 3.8. Published in Albert Lehninger's biochemistry textbook, the figure presents an enzyme reaction by way of a sequence of chemical formulae. The representation is accompanied by text blocks describing the changes that accompany the different stages of reaction. One can follow how the reactive group of the enzyme, the rest of which is only schematically hinted at, interacts with the substrate, and which transition processes occur in the course of the reaction. All in all, we are still caught in the representational world that was adopted from chemistry in the second half of the nineteenth and the first half of the twentieth century and which dominated the teaching literature until the mid-twentieth century. It

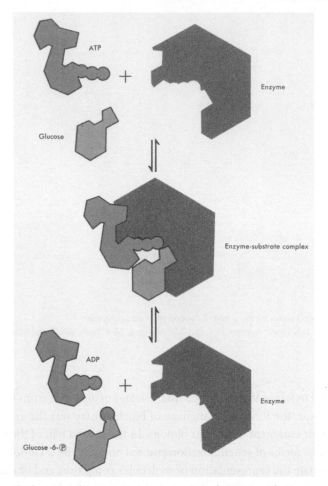

F I G U R E 3.9. The formation of an enzyme-substrate complex with ensuing catalysis
Source: James D. Watson, *Molecular Biology of the Gene* (New York and Amsterdam: Benjamin, 1965), 52,
fig. 2-8.

was actually a replacement for an earlier, even more text-oriented literature.[14]
Beyond the structural formulae, there are no further figurative components
as yet, but the process is made more legible by the omission of those parts of
the formulae that are not relevant to the partial reaction in question.

Figure 3.9 stems from the textbook by Watson mentioned above. Although
written around the same time as Lehninger's textbook, Watson's book brought
about something like a paradigm shift in the forms of representation of molecular biology, demarcating it not only conceptually but also pictorially from
classical biochemistry.[15] Besides electron micrographic representations, the
textbook abounded in graphic schematizations. The figure also shows an en-

zyme reaction, this time an anonymized one. At first glance, we can see that the close fit of the participating substances—enzyme and substrate—is presented according to a lock-and-key principle, while the contact of the enzyme with the substrate provokes a change in conformation: an induced fit. What we find in the picture are no longer molecules as atomic associations, but the scheme of a stereochemical principle. Formulae have completely disappeared from the image, and it manages with an absolute minimum of designations. It is certainly of interest that the idea of a lock-and-key fit dates back to the late nineteenth century.[16] Its schematic representation did not, however, gain acceptance under the old formulaic regime, and caught on in biochemistry

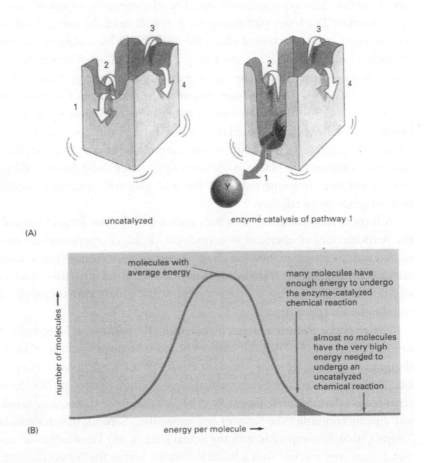

FIGURE 3.10. Enzyme catalysis
Source: Bruce Alberts et al., *Molecular Biology of the Cell*, 3rd edition (New York and London: Garland Publishing, 1994), 64, figs. 2–16.

$$E + S \rightarrow [ES] \rightarrow E + P$$

FIGURE 3.11. Schematic formula of an enzyme reaction

and molecular biology only about half a century later, although it was already accepted knowledge in immunology around 1900.[17]

Yet another representation of enzyme catalysis is to be found in the successor textbook to Watson's *Molecular Biology of the Gene*, the *Molecular Biology of the Cell*. This representation visualizes the energetic aspect of an enzyme reaction. The lower part shows the distribution of the energy levels of biomolecules, only a fraction of which are amenable to bio-catalysis, whereas the bulk of molecules are not. And only a small number of molecules are able to react spontaneously. The upper part shows the energy barrier to be overcome for a number of reactions, as well as the lowering of that barrier by enzyme catalysis for one of these pathways as a box with outlets at different heights. The schematization used here is interesting in its combination of a completely intuitive drawing with a semi-quantitative graph. Although these two representations are placed in the same figure, they show different things that are not directly connected at all. But both play with qualitative figurations of quantitative relations.

A form of schematization that goes even a step further forgoes not only the particularities of chemical structure and all the energetic aspects connected with the reaction, but also all stereochemical imagery. What remains is a reaction sequence—not to be mistaken for a chemical formula—in which enzyme, substrate, enzyme-substrate complex, and product are simply designated by letters (fig. 3.11).

The physicist Heinrich Hertz, a student of Hermann von Helmholtz in Berlin, once contended that the sciences produce "internal simulacra" of the things of the world, whereby "different images of the same objects" are possible.[18] He postulated three requirements that such images must fulfill to be regarded as scientific. First, they have to be "admissible"—that is, they should not directly contradict the "laws of our reasoning." Second, they have to be "correct," that is, compatible with the actual state of our knowledge—a state that is, however, marked with a historical index, just as the "correct" images are. Third, they should be "useful," a requirement that for Hertz basically meant that they should follow an economy principle of representation as defined by Ernst Mach.[19] On this latter level, he saw conventions prevailing.

To summarize: the visualization modes of condensation and of dilatation discussed in the first section of this chapter are frequently bound to the experimental setup and the research technologies involved, and essentially predetermined by them. The mode of enhancement, by contrast, reveals itself to be closely linked to a particular class of epistemic objects, the preparations. The visualization mode of schematization, in turn, can be linked to models, preferably with models of function and of mechanisms. Schematizations are also particularly apt for translocating the knowledge objects in question into the circulation and transmission spheres of knowledge, respectively. Visualization thus permeates not only the totality of the space of experimentation, but scientific discourse as a whole.

The review of methods of visualization presented here certainly makes no claim to be complete. Moreover, the different procedures described do not, as a rule, occur in complete isolation from each other. Rather, they refer to each other in any given context of visualization; they support and complement each other with their respective particularities. In addition, it is quite common that all three of the described procedures of visualization, either combined or distributed over a number of different techniques, accompany the text of a single original publication. To the extent that the principle of visualization has become the dominant mode of reports in the natural sciences, it is arguably now the text that accompanies a shorter or longer series of visual representations. They form mutually referential chains or networks of representation. And insofar as they are based on or derived from techniques that do not depend on each other, they can be taken as independent pieces of evidence for the phenomenon in question. The better they relate to each other, the more stable a finding appears to be. A further possibility is that one form of visual representation becomes the raw material for another. In this way, the objects of research are brought—literally—to manifest themselves in the space of traces and data. These manifestations are essentially inseparable from the techniques to which they owe their existence. This is the center around which a phenomenology of experimentation revolves. It is one of the principal concerns of an epistemology from below to preserve the research technologies behind the representations that derive from them.

4

Grafting

The effects of generating traces, modeling, and visualization we have discussed so far belonged more to the epistemic side of the process of experimentation. In this chapter we shall look more closely at its technical side: aspects of the procedure and the omnipresent techniques that enable experimental work in the first place. In particular, this concerns the insertion of new apparatus or procedures in already existing experimental setups. An experimental environment is seldom formed from scratch. In most cases, systems already in use undergo a transformation by apposition. This is, however, a delicate operation, often requiring a considerable degree of adjustment and refinement, for it has to meet two requirements. First, the implementation should conserve the actual state of the art; and second (and prospectively), it should make it possible to achieve things that were beyond the reach of the existing state of the system and its limits. I call this procedure grafting. I will embed these considerations on experimental grafting in a broader context of cultural techniques of expansion and propagation.

Grafting as a Cultivating Technique in Biology

I begin by looking at the *biological* techniques of culturing by grafting and related procedures. We have to ascertain the peculiar and distinct quality of these techniques to be able to assess what grafting in experimental systems can mean and which forms it can assume. In addition, we have to clarify the relationship between the technique of grafting and neighboring techniques such as vaccination, transplantation, and hybridization as the many different forms of organic allomorphy.[1] It goes without saying that in this field we are dealing with practices that developed and changed over long periods of time.

Accordingly, we have to keep in mind the semantic dislocations depending on the temporal succession of the technologies that have determined and still determine the field today. Such a preliminary digression is at least worthwhile, since many of these concepts are used productively in the cultural as well as the natural sciences today, generally without awareness of their origins. However, instead of the usual practice of describing such displacements in the vocabulary of metaphor or analogy, in this case I shall follow Isabelle Stengers and envisage them as migratory movements that promote knowledge, and the participating concepts as ambulatory.[2]

Grafting—an expression used in pomology in particular—indicates a specific form of vegetative propagation. The technique of grafting is applied mostly to fruit-bearing trees, but it is also widespread in the propagation of ornamental plants. There are two reasons why it can be desired, if not necessary: either the fruits are the result of a hybridization and the plants would, if sexually propagated, revert for the most part to their pre-hybrid forms; or the fruits are no longer capable of being propagated by their seeds because they have degenerated genetically as a result of their breeding history. Grafting is also of interest when the aim is to combine the properties of the stock with that of the sprig. The principle is as follows: a twig or a bud of the desired variety is applied to a congeneric but not identical stock in such a way that both parts grow together. This can be effected by whip grafting, notching, or "budding onto the sleeping eye,"[3] as it had been called, among other terms of the art. The latter technique is described by the well-known nineteenth-century French cultivator Alexis Lepère from Montreuil in his book on the culture of peaches (fig. 4.1). Traditionally, one speaks in this context of refinement, but strictly speaking, this expression is misleading. The whole thrust of grafting consists of both parts keeping the respective properties that led to them being brought together.

The following example may make the point. Toward the end of the nineteenth century, the roots of the traditional European vine varieties were stricken by the grapevine louse *Phylloxera vastatrix*, which was accidentally imported from America and proceeded to destroy the vine stocks of many European countries on a huge scale.[4] The endemic American vines, however, were immune to attacks by the insect. By grafting the European cultivars onto American stocks, the new stock's resistance to the pest persisted, while the European sprigs that were grafted onto the root stocks still yielded the traditional Pinot Noir or Riesling. Despite a tight intergrowth at the point of intersection that usually remains visible, the parts retain their specific genetic identity upon grafting, even if the stock can influence the growth of the graft to be more or less vigorous, and its twigs to bear more or less fruit.[5] As far as their genomic

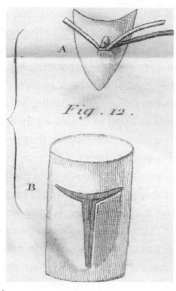

FIGURE 4.1. Budding of peach trees
Source: Alexis Lepère, *Pratique raisonnée de la taille du pêcher* (Paris: Bouchard-Huzard, 1846), plate I,
fig. 12.

properties are concerned, however, they remain heteronomic with respect to
each other. The cultivation technology of grafting already advanced this genetic tessellation in antiquity, long before their cellular basis was elucidated.

Grafting is thus a peculiar connection. The concrescence is complete, and
sometimes even no longer perceptible. However, it does not mediate or mix
anything; on the contrary, it leaves the parts their respective identity, even
though the graft uses the stock for its own growth, and the stock profits, for
example, from the photosynthetic activity of the graft.

To highlight the peculiarity of grafting in contrast to other methods, let
us compare it with a series of further biological technologies of transmission.
It is remarkable and not by chance that those technologies not only play an
important role in medicine but are also an integral part of biological experimentation. In recent decades they have, in addition, been taken up by science
and technology as well as media and literary studies to highlight processes
of transmission in other areas of cultural practice. I will not delve deeper
into the literature here, save to mention some prominent texts in passing. I
begin with vaccination, or inoculation. The notion of inoculation was, for a
long time, used more or less synonymously with the notion of grafting. In the
late nineteenth century, however, it definitely changed its meaning.[6] Vaccination today means the purposeful application of a pathogen that has been deprived of its virulence, or parts of it, to a higher organism. This stimulates the

immune system of the recipient to protect it against the virulent form of the germ. We are therefore dealing with the transient intrusion of one organism into another, by which the latter, as a reaction, changes one of its properties in a relatively durable manner, to become resistant against future attacks of the disease agent. Vaccination is thus qualitatively and essentially different from grafting. In the case of the vine, the resistance was bestowed by the new root and not induced by the graft. However, in the case of vaccination, as it is understood and practiced today, the latter is the case: the attenuated pathogen induces the inoculated organism to change by itself, rendering it immune to the particular disease-causing agent.

A third biological cultivation technique of transmission is transplantation.[7] In today's experimental practice in medicine and biology, transplantation is understood as implanting a part, typically an organ, of one individual into another, and letting it become ingrown. The aim of a medical transplantation is usually to replace an organ or tissue that has stopped functioning properly with a functioning one that is as similar as possible in its vital state to the organ to be replaced. Ideally, after transplantation, the organism does not change as compared with its previous, healthy state. While grafting serves the purpose of transmitting a part with a new property to an organism and letting it grow in permanently, transplantation aims at keeping the intact state of a given property identical while exchanging a part.

The fourth biological cultivation technique to consider is hybridization.[8] Again, this differs significantly from the procedures already described. Hybridization—or bastardization—means that two strains of an organism that differ in a significant character but can propagate sexually are mated with each other. The aim is usually to combine the characters of the two parents in the offspring and, if several different characters are in play, to create an organism that is qualitatively different from either of the two parents. In contrast to grafting, it would therefore make sense to talk about refinement in the case of hybridization. Think of an apple that, through crossbreeding, is tasty because it has the right proportion of sweetness and acidity, and can also be kept, for instance, throughout the winter period. With the establishment of classical genetics, hybridization was imported from agriculture into science and transformed into an experimental technology with which it became possible to analyze hereditary processes and genetic constitutions.[9] And although this meant that the term shifted its meaning, the hybridization of nucleic acids, in the framework of molecular genetics, advanced to become one of the decisive experimental techniques for the elucidation of the molecular structure and function of the genetic material.[10] We can see that not only concepts, but also related procedures, can migrate from one area of culture to another. To speak here simply of metaphors or analogies would in no way do justice to these historical movements. Migrations of this kind have an eminent knowledge-generating potential. We shall now consider the technology of grafting in more detail, first in the context of a poietics of writing and subsequently in the context of scientific experimentation.

The Graft of Writing

All the culturing techniques listed here are processes of manipulation by which an exchange or a new connection occurs between two separate and independent entities. In a general sense they are "operations of propagation" and "operations of transition." These are the expressions Isabelle Stengers used to describe and analyze conceptual transactions within a particular area of knowledge, and for conceptual migrations between different disciplines and scientific fields.[11] Such movements, which Stengers also calls "nomadic," are a prerequisite for cultural techniques characteristic of widely different practices to shed light on each other. These practices range from manipulating organisms to practices of writing, from artistic to scientific practices of experimentation.[12]

In relation to the transposition of linguistic particles from one discursive context into another, Jacques Derrida has spoken of an—insatiable—"iteration."[13] In referring to the linguistic world of biology, he also used concepts such as "dissemination" and "grafting"[14] to characterize the play with the heteroclite, the heteronomous, and the heterogenous that he saw as the decisive mark of the cultural technique of writing, as intrinsically giving rise to a permanent *Fortschreibung*.[15] Dissemination appears as a particular operation of propagation comparable to an illness spreading through the body in a more or less erratic way, and is therefore not linear in character and cannot be predicted exactly. In contrast, grafting means the insertion of something foreign that does not organically emerge from the preexisting structure, although it requires it in order to articulate itself as that other. For Derrida, dissemination and grafting are both figures of spatial and temporal propagation in the field of writing and the written.

Grafting comprises more than just citation as a special scriptural procedure of insertion. "To write means to graft," Derrida asserts categorically and succinctly in *Dissemination*, in the sense of applying something to a support as well as of the decontextualization and recontextualization of set pieces. And it is here that Derrida associates grafting with the application of an "*overcast seam*" (*surjet*).[16] This means that grafting implies a contact surface on which a durable conjunction is established as a suture, which he distinguishes from a fusion.

In his book *Water and Dreams*, Gaston Bachelard also utilized the concept of "graft," this time in the context of a characterization of poetic work on and with images of the elements—fire, water, earth, and air—to express the patterns of transposition and the motion sequences of poetic images in the process of writing.[17] In the context of his "philosophical study" of "poetic creation,"[18] Bachelard emphasizes that the graft adds something new to the stock, something that cannot be deduced from it in the sense of an imaginative derivation. And yet the graft is dependent on the support in order to be able to sprout its own foibles. The "material imagination" is the result of an apposition. With respect to the "images that stem *directly from matter*," Bachelard writes: "The eye assigns them names, but only the hand truly knows them."[19] And elsewhere: "Every poetics must accept components of material essence."[20] It is obvious that Bachelard is expressly refusing to understand the "graft" as a "mere metaphor." For him, it is rather a figure of the movement of "material imagination." In other words, he takes it at face value. Material imagination is essentially event-like: more than a mere expression of the power of imagination, it is driven by a vegetative force, itself matter-like and at the same time the result of an engagement with matter, an engagement that surpasses itself. "It is the graft which can truly provide the material imagination with an exuberance of forms."[21] Exuberance is a fitting description of the kind of surplus we are dealing with here. "A written work requires, in order to be more than a way of simply filling in time, it must discover its *matter*."[22] And although in this context Bachelard is occupied primarily with poetics (that is, the literary work of art), he adds in a generalizing and indeed categorical fashion: "Art is grafted nature."[23] It is not that it is made from one piece; instead, its existence relies on the combination as well as the displacement, the match and mismatch of the specific materials that go into it, with their own distinctive and irreplaceable intransigence.

In the written work, the imaginative material and the linguistic material in a narrower sense—semantics and syntax—are interlocked; they are intertwined with one another. In writing, as in other creative processes, one does not just bring something intrinsically coherent into being. One permanently adds from outside. In writing, one is taken and subverted by the formulations. This figure of subversion is due to the material dispositive of writing. It is an experimental arrangement, a practical "generative disposition" in the sense of Pierre Bourdieu,[24] a culture allowing for "the access to an emergence" in the sense of Bachelard.[25] One can get caught up in its maelstrom. With things that are written down one can differentially work further, can practice apposition, and poetic as well as knowledge effects happen through repetition. They need writing in order to be generated; they only happen in its course. What can be said here in agreement with Bachelard about the literary text also applies *a fortiori* to scribbling in the space of experimental note-taking. Here, too, one grafts, bringing previously unconnected things into preliminary relation. Chapter 5 will deal with experimental protocols in more detail.[26]

Grafting in Experimental Systems

This book is about the modes and shapes of articulation and the dynamics that characterize the phenomenon of scientific experimentation in its material constitution. As we shall see, here there are also good reasons to use the concepts of grafting and hybridization. The difference between the two cultivation techniques in the biological realm can be formulated briefly as follows: a hybrid is characterized by the amalgamation of two previously separate entities, whereas a graft, although it grows together with its support, allows both parts to retain their identity. Both techniques have their equivalent in the field of experimental technologies. Experimental techniques are a particular form of cultural technique in the realm of knowledge production. We have to be careful not to confuse them with method as it is generally understood. A method is usually defined as a self-consistent instruction to act that can be *applied* to an existing problem. Research experimentation, in contrast, is about a frequently antinomic combination of heteroclite research technologies into precarious and fragile ensembles that are able to function conjointly as an experimental system.

Claude Lévi-Strauss referred to the tinkerer's characteristic of drawing on a repertoire whose "composition" is "heteroclite."[27] He saw the same principle of composition at work in "wild thought." He contrasted it with the mindset of an engineer—not a scientist—whose work he saw as subject to a unifying discipline. As we shall see, a bricolage, wild thinking, is also at work in the deepest recesses of the scientific research process—in the innermost realms of a science in the making. I will come back to this point in chapter 9. The problem is to find a vocabulary that allows us to capture such tinkering in its phenomenal articulations, and to represent those articulations in their dynamic action.

Experimental systems can be seen as the smallest functional units of the modern empirical research process. And it is important to recognize that, in essence, they are temporally constituted: temporality is inscribed into them. They are engaged in a permanent process of differential reproduction—again a concept that refers to the realm of the organic, but has also acquired a central importance in economy. Experimental systems must differentiate themselves continually if they are to retain their function as generators of new knowledge. Expressed more neutrally, without equating differentiation with increasing complexity and consequently ascribing a direction to it: an experimental system must always differ—it has to "iterate" steadily.[28] As soon as such a system settles into itself, it loses its research function and is only capable of self-demonstration—as a test. We are therefore dealing with systems of a very particular kind. They are laid out not for a maximum of coherence, but for a minimum of cohesion.

In order to avoid settling into itself and opting out of the research front, the technical conditions of an experimental system are permanently reworked and tinkered with. Experimental systems revolve around epistemic things, objects of knowledge in the process of gaining shape. New elements can be flanged to the technical conditions under which an epistemic thing can acquire contours. If we are concerned not just with the modification or incremental alteration of an already existing element, but rather with the introduction of a qualitatively new apparatus into the system, we can rightly compare it with the procedure of grafting. Even if we are dealing not with an organic but with an epistemic-technical ensemble, it nevertheless satisfies formally comparable requirements. With the help of the support, the graft will have to unfold its own potential, while simultaneously leaving the functioning of its support intact. What is more, as we shall see shortly, in experimental systems that hold entities from the realm of the biological at their center, the surrounding research technologies are compelled to interact with these organic entities in a highly characteristic manner. They form precarious, organic-inorganic interfaces with an enabling character. They have to be designed not to monopolize the object but to help it manifest itself in its intrinsic value, in the sense suggested by the early media theorist Fritz Heider when he stated that "in the physical structure itself—without reference to a particular subject—differences are in place that predetermine certain things as mediators, and others as objects."[29]

New technical devices can thus be grafted onto existing experimental systems. A few examples will elucidate the features of this process. In *Toward a History of Epistemic Things*, I reported in detail on the history of the experimental system of in vitro protein synthesis. Around the middle of the 1950s a new instrument, the ultracentrifuge, was introduced. It replaced traditional laboratory centrifuges and enabled a leap toward significantly higher sedimentation velocities, which led to qualitative changes in the system.[30] It generated a new fractionation pattern of the cell sap being observed, and it led to the identification of previously unknown molecular components which were found to play a crucial role in the cellular mechanism of protein synthesis. Figure 4.2 is taken from a review article Mahlon Hoagland, who played a

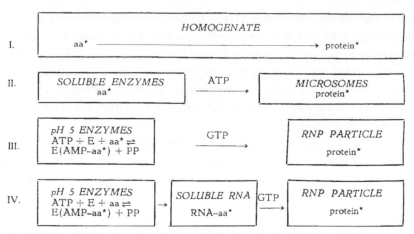

FIGURE 4.2. Stages of the dissection of the system of cell-free amino acid incorporation into proteins on the basis of rat liver
Source: Mahlon Hoagland, "On an Enzymatic Reaction Between Amino Acids and Nucleic Acid and Its Possible Role in Protein Synthesis," *Recueil des Travaux Chimiques des Pays-Bas et de la Belgique* 77 (1958): 623–33, fig. 1.

leading role in this work, presented as a summary and retrospective toward the end of the 1950s. In historical stages, it shows the changes resulting from the insertion of this technique with respect to the differentiation of the components of the system and the epistemic object that was the focus of the investigation. The background for the diversification of the molecular components in the boxes of the graphic is the refinement of the fractions. Up to stage II, a standard laboratory centrifuge was in operation; from stage III onward, a high-speed ultracentrifuge was added. This shows neatly how an equipment-related graft that at first looks more like a linear and quantitative extension of the existing experimental conditions—the increase of the centrifugation velocity—changes from an incremental differentiation to a qualitative transformation of a research object. In particular, it became possible to localize the energy requirement of the process. What emerged was a completely new research object that is still tagged here as "soluble RNA" and was later renamed transfer RNA.

Our second example demonstrates how different the impact of such grafts can be. In the late 1930s, a biochemical procedure of analysis was introduced into the system for the investigation of physiological gene actions by the zoologist Alfred Kühn at Berlin's Kaiser Wilhelm Institute for Biology. The system was based on mutants of the flour moth *Ephestia kühniella*; I have discussed it in detail elsewhere.[31] Until then the investigative procedures had essentially been based on the transplantation of organs between wild type and eye pigment mutants of the insect. In contrast to medically motivated transplantation, the use of these organs as an experimental technique is understood not as a *replacement*, but as a *relocation*. In the case at stake now, an additional procedure—biochemical analysis—was integrated into the experimental ensemble instead of extending a technology—centrifugation—that was already in operation, as in the previous example. With this move, the group around Kühn was able to characterize the gene action substance as the molecular derivative of an amino acid that had previously been addressed as a putative hormone.

Following this finding, a distinction could be drawn that became seminal for future research: Kühn and his coworkers could now differentiate between two "action chains" in gene physiology. One of them concerned the putative genes, with their associated ferments (enzymes), as the assumed direct gene products. The other involved the physiological substrates that were transformed by the enzymes in cascade-like steps (fig. 4.3). Here again, we can see how a technical graft results in the iterative differentiation of an object of investigation. The decisive step from a one-dimensional, linear representation of gene action to a two-dimensional, reticulated representation was once more the result of a technical implementation. This tight coupling of

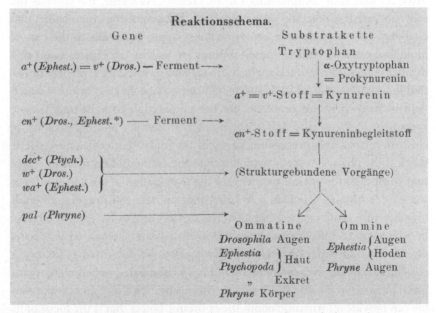

FIGURE 4.3. Reaction scheme of the relation between gene action chains and substrate chains
Source: Alfred Kühn, "Über eine Gen-Wirkkette der Pigmentbildung bei Insekten," *Nachrichten der Akademie der Wissenschaften in Göttingen, Mathematisch-Physikalische Klasse* (1941): 231–61, esp. 255.

technical and epistemic things is characteristic of the modern research process. The graft should not only augment the technical system; from an epistemic viewpoint it must also deliver a result that is characteristic for that augmentation. Otherwise, it would not be worth the effort.

Interfaces

Yet another concept comes into play here that has already been referred to in passing: the concept of interface. In connection with grafting in experimental systems, interfaces assume a twofold form. On the one hand, we have the seam between the grafted new technique and the techniques already in place—that is, the technical conditions determining the experimental system. The graft must be compatible with the existing technical system. It must be able to connect to it in such a way that its effects can unfold. It must be possible to relate the new traces and data being generated with its help to the existent data and traces. If such a compatibility is not given, the grafting procedure has to be judged a failure: the system rejected the graft.

On the other hand, we have to observe the interface between the new instrument or procedure and its target, the epistemic object. Here, too, productive compatibility has to be achieved—there must be at least a reasonable prospect that such compatibility can be created. As a rule, the latter is the case, because it is precisely such precarious, but promising connections that challenge the tinkering with experimental systems. Especially in the case of biological research objects, these jointings are usually the critical point that decides the possibility of the application of a new technology. It often happens that the configuration of such interfaces itself widens into a genuine research field of its own. Good examples are the preparation of light-microscopic specimens in the second half of the nineteenth century and the creation of electron microscopic preparations around the mid-twentieth century. Here, it is less the physics of the instrument than the configuration of its interface that constitutes the critical moment for the exploitation of the capabilities of the new technique. The task is to bring more or less soft organic materials together with more or less rigid technical devices and to have them interact in such a way that the contact leads to traces that can be regarded as specific for the material under observation. The goal can be described as follows: a conductive contact area or contact point has to be created between an organism—or a part of it—and a technical apparatus in such a way that the apparatus can provide information about the epistemic object that is the focus of its analytic differentiation potential. Such interfaces can assume widely different

forms, as I have discussed elsewhere in more detail.[32] In physiological experiments with living animals such as those carried out in the second half of the
nineteenth century, a tube filled with mercury might well have served as a
connecting link; in the case of the analysis of biological macromolecules by
X-ray diffraction, the interface assumes the shape of an organic crystal. A
typology of such interfaces would be instructive, but remains a desideratum
to this day.

Hybridization

Finally, I would like to talk briefly about yet another articulation of experimental systems: hybridization. If grafting moves within the technical frame
of a single experimental system in development, hybridization stands for the
fusion into a new construct of two such systems that were established independent of each other. Complementarily, we can also observe the ramification
of a single experimental system into two or more unit-systems that specialize
in the representation of sufficiently different epistemic objects. It is not this
latter form of dissemination, but its inverse complement, hybridization, that
we shall focus on in the last section of this chapter. In most cases, the result
of such melding is the emergence of a new scientific object of investigation.
Sometimes the hybridization of experimental systems leads to the opening of
completely new research fields. The merging of previously separate research
trajectories is one of the richest sources of scientific innovation.

A textbook case of such a hybridization process in the history of molecular biology is the fusion of Jacques Monod's experimental system for the
investigation of the regulation of bacterial sugar metabolism—lactose breakdown, in the event—with François Jacob's system for the investigation of gene
transfer in the sexual conjugation of bacteria.[33] During the 1950s at the Pasteur Institute in Paris, both researchers used the same model organism, the
bacterium *Escherichia coli*, but each of them started out by pursuing a completely different project. Monod was interested in how bacteria react physi-
ologically to various nutrition sources by producing the enzymes necessary
for their breakdown, or ceasing production of momentarily superfluous enzymes. Jacob was immersed in studying the phenomenon of lysogeny in the
laboratory of André Lwoff: lysogenous viruses that have infested a bacterium
can eventually become integrated into the host genome and can be carried
along over many generations in a quasi-dormant mode before they become
virulent and destroy the bacterial cell. The fusion of the two systems became possible because both researchers based their work in the same model

organism. By way of differential gene transfer between genetically different strains of bacteria, Monod and Jacob, in collaboration with Arthur Pardee and Monica Riley from Berkeley, succeeded in mapping the genes responsible for lactose breakdown on the bacterial chromosome and developing a model of its regulation. They also reached the conclusion—it became one of the cornerstones of molecular genetics—that gene expression is generally mediated by a more or less short-lived intermediate, a ribonucleic acid derived from the information-carrying DNA that was soon to become known and to enter the textbooks as messenger RNA. As a result of this hybridization of two experimental systems, three new epistemic objects appeared on the scene of fledgling molecular biology: the distinction between structural and regulatory genes, the operon as a unit of bacterial gene regulation, and messenger RNA as the mediator of gene expression. In parallel to the hybridization of the experimental systems, a hybridization of the experimental discourse occurred.[34] The multilayered character of the hybrid language that emerged from this fusion then helped in the exploration of further modulations of the newly constituted experimental system.

The two forms of articulation described here—grafting and hybridization—play a decisive role in the iteration of experimental systems. Grafting concerns the extension of experimental systems by new research technologies. Hybridization concerns the fusion of formerly separated experimental systems. Both forms of articulation are sites of tinkering at which unexpected effects can happen, resulting in equally unexpected additions to knowledge. This characterization is aimed not so much at introducing a biotechnical vocabulary into the description of the dynamics of modern empirical research as at directing attention to the fact that, whether here or there, we are dealing with cultural techniques that can occasionally even literally migrate from one area to the other. Its connotation is material, not linguistic.

Regarding the description of Alfred Kühn's experimental system, we have seen that an early technique that was integrated into the system consisted of the crosswise transplantation of insect organs into the caterpillars of different strains. In this case the medical practice of transplantation becomes a component of the experimental arsenal of procedures of genetic research, and can have consequences and be subject to reinterpretations that would be excluded in the traditional context. Similar things happen in the context of hybridization. The bastardization of strains and stocks of a species (a technique long known and practiced in agriculture) became, alongside the breeding of pure lines, the decisive experimental technique for the establishment of classical genetics in the period around the end of the nineteenth and the beginning of the twentieth century.[35] In the scientific research context, however, the manipulation of biological reproduction or of other biological functions for cultivation purposes is not at issue. What counts here are procedures of knowledge production. In the biological sciences, however, many of these procedures have a long history of cultivating and manipulating animals and plants. As procedures of experimentation, they have now acquired their own potential for iteration.

5

Protocols

Drawing up protocols is an integral part of any experiment. In the foregoing chapters on traces and models, and on visualization, the act of protocolling has been mentioned repeatedly, if at times only implicitly. In what follows I will try to explicate this form of writing, which derives its meaning from its vicinity to the experiment, and to capture and locate its various facets more precisely and explicitly. These primary written notes and jottings extend over a phenomenally large and multifarious area. They range from the design of an experiment and the listing of its results to excerpts of the research literature, notes on basic conceptions, fragments of an idea, intuitions, remarks on conspicuous overlaps, sketches of experimental arrangements, stripes of data derived from a particular experimental run, tentative interpretations of experimental results and revisions of such interpretations, preliminary calculations, calibration results of apparatus to be utilized, and drafts of new contraptions. All these preliminary notes are a permanent accompaniment for the experimenter. They are located in the space between the materialities of the experimental setup and the conceptual and narrative structures that ultimately leave the immediate context of the laboratory and enter the sphere where knowledge circulates.[1]

Although orderly experimentation is impossible to imagine without these forms of registration, for a long time the primary means of scoring and chalking up in the laboratory were seen simply as necessary to fix and record data, as storage on the way to the definitive result. They consequently received little attention from epistemology. This situation has only changed recently.[2] The types of written temporary storage are by no means the inert scaffold for the final representation of a result. They are, rather, a productive medium between the objects of experimentation and the knowledge events connected with them. The laboratory is a space in which new knowledge arises; in this context, hastily written rough notes, jottings on slips of paper, and full and rewritten protocols mediate the concrete processes of knowledge formation. At that point of the experimental process, there is no authoritative voice that would already have decided where to go. The tentative prevails. The *productive* function of such documents in the process of knowledge generation has, however, been neglected in most of the microhistorical reconstructions based on laboratory notebooks and other documentation—and remnants—of research.[3] They have mostly been concerned with closing gaps in reconstruction left by the published documents—that is (to quote an expression used by Frederic Holmes), with the fine mapping of "investigative pathways,"[4] the formal modeling of discovery processes,[5] or the provision of resources for historical research.[6]

In principle, two forms—and phases—are to be discerned to which we can assign laboratory records, the kinds of notes made near to the scientists' workbench. On the one hand, these are notes that serve to prepare an experiment. They can assume the form of more or less cursory, rough jottings, but they can also appear in the shape of partly formatted schemata. On the other hand, we have the records and annotations that are made in the context of evaluating an experiment. The former help to prepare, the latter to process the experiment. What follows in this chapter is devoted to these two kinds of notes. Until well into the beginning of the desktop computer age, such proleptic sketches and the subsequent evaluations were predominantly handwritten on paper, usually in bound notebooks or in folders. Early on Peter Medawar pointed to the importance of notes and laboratory protocols to an adequate understanding of scientific work.[7] With the continuing and ever-increasing digital provision of masses of data, their processing has relocated and the results are now stored in the computer. This raises the inevitable question of whether something basic has changed in the system of written records close to the process of experimentation. I will return to this question in the last section of this chapter.

Notation

To begin with, I shall consider the textualizations in the run-up to the experiment. Here, the classical function of the notebook comes into play.[8] In intense phases of experimentation, experimenters are like prisoners of the procedure. What is at stake in scientific research processes is the generation of unprecedented knowledge. We are dealing not only with consequences that can be derived from the knowledge that is already there, but first and foremost with effects of knowledge that cannot be anticipated. In a sufficiently complex experimental process, aspects of an investigated phenomenon can become manifest which the experimenter could not have envisaged. Everything revolves around these experimental contingencies. The art of experimenting consists in reacting appropriately to these things. A person who gets involved in this transaction requires a double prerequisite that seems paradoxical at first sight. One of the prerequisites is familiarity with the material. What matters here is the point at which experimenters have acquired such an intimacy with the matter in hand that they have the feeling of being able to act from inside it. At the same time, however, this condition enables the experimenter to bestow sovereignty on the matter at hand. To use a well-known topological image: as with a Möbius strip, there is, on the one hand, a clear and visible distinction between inside and outside, between front and back, between subject and object; and on the other hand, there is no point at which a borderline can be drawn across which the two would be clearly and distinctly opposed.

This kind of vicinity in distance and distance in vicinity that characterizes the dispositive of the experiment involves a certain type of anticipatory attention on the part of the experimenters. They are seized by a sort of permanent frenzy accompanying not only the experimental process itself but also their everyday actions and behavior. This is exactly where the topological point of the note lies. For a long time, and presumably still today, the pages of a handsome notebook were the privileged place where a researcher could hold fast not only daily trivia such as shopping lists, addresses, and other things, but also, and more importantly, the ruminations, intuitions, and possible twists and turns and trains of thought involved in an experiment in progress. Ernst Mach's notebook no. 25 is a good example. Its pages were filled between February and April 1887 in the context of Mach's and Peter Salcher's photographs of bullets, which have been analyzed in detail by Christoph Hoffmann and Peter Berz.[9] As a rule, such notes appear cryptic at first sight, elliptic, decontextualized, and therefore not easily integrated into the stream of the experimental events; but they can provide decisive information—in particular on unexpected turns in the experimental process. Sometimes, however, such entries can also be traced to extraneous events such as changes of location. This is precisely what happened to Mach at the beginning of his stay with Peter Salcher in Fiume when he observed the undulation caused by a ship in the sea and sketched it on paper.[10]

A notebook of a somewhat different character is Claude Bernard's *Cahier rouge* from the mid-nineteenth century.[11] Along with his experimental laboratory books, this notebook documented the laboratory work of the French physiologist between 1850 and 1860—that is, during the most productive years of his experimental activity at the Sorbonne and at the Collège de France. This notebook is full of thoughts—often including preliminary sketches— about subsequent experiments that crossed Bernard's mind when he was ex- perimenting on similar topics. Here, too, no systematic order can be observed as organizing the entries. Mirko Grmek, to whom we owe the critical edition of the laboratory booklet, cautiously describes "a certain absence of order."[12] The thoughts on future experiments follow each other without being directly related. Frequently, however, Bernard left space on the pages for further en- tries, and in this way he could return later with remarks on earlier proposals. The notes are of an "apparently heteroclite character," as the editor comments.[13] Mostly they take the form of a question and concern either experimental con- trivances to be constructed or animal organs to be prepared. In between are frequent insertions concerning general ideas about experimentation as well as basic remarks on the science of physiology. Grmek states, "We are con- fronted with a mind that is about forming itself, not with a fully accomplished mind"; and elsewhere, "[Bernard] sometimes makes use of contradictory as- sumptions, hesitates, does not decide."[14] The notebook attests to an experi- mental regime that offers more options than could ever have been realized under Bernard's direction, and it foreshadows Bernard's synthetic and episte- mological as well as general physiological late work, notably his *Introduction to Experimental Medicine* of 1865[15] and his *Lectures on the Phenomena of Life Common to Animals and Plants* (1878–1879).[16]

Experiment Sheets

An experiment rarely comes alone. I will return to this issue more extensively in chapter 6. Series of experiments can extend over a period of years and follow a scheme with incremental modifications. Such temporal stretching has literal consequences. It gives rise to a genre of paperwork that can be considered as half-standardized. These are schemata that, first, prescribe a scaffold of recur- ring steps in the course of an experiment, and second, allow that sequence to be filled out flexibly according to the individual orientation of each experi- ment. This category of collective notational techniques comprises forms that follow a protocol of successive steps tailored to a particular experimental sys- tem; as a rule, they do not leave the rooms of a laboratory but are shared collec- tively inside it. It also comprises laboratory manuals that can leave their mark on a whole laboratory culture (more about such cultures in chapter 7). These forms of written notation can be seen as literal experimental technologies.

Let us dwell for a moment on these experimental flow protocols that ac- company the laboratory work. Figure 5.1 shows such a semi-codified working

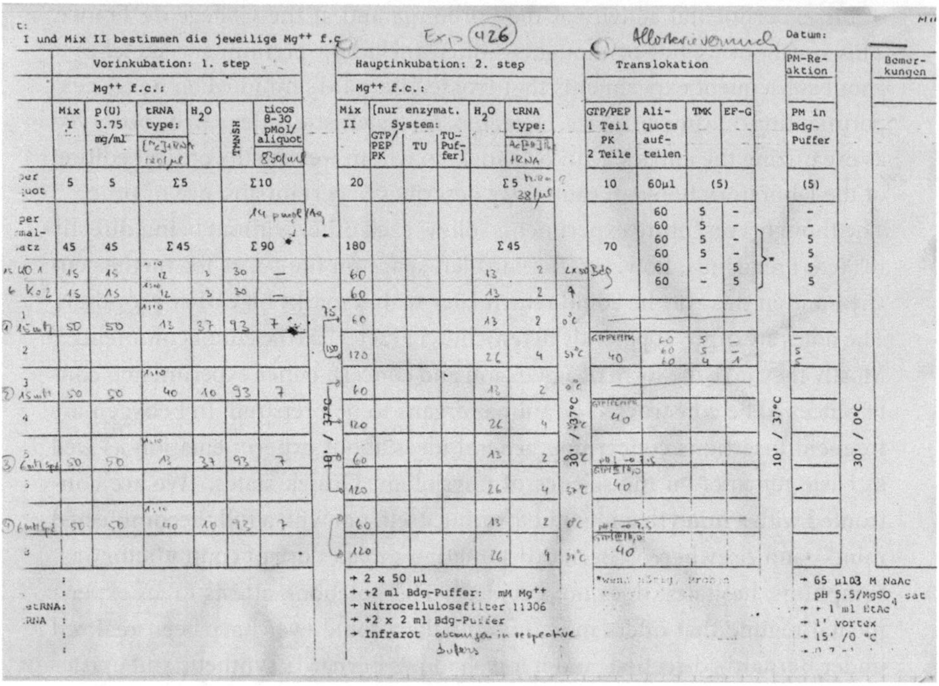

FIGURE 5.1. "Allostery assay," experiment no. 426, flow scheme
Source: Archive of the author.

scheme. It is the layout of an in vitro experiment for the molecular exploration of protein biosynthesis from a laboratory of the Max Planck Institute for
Molecular Genetics in Berlin in the 1980s. As we can see, it is a hectographed
schematic handout that has been filled, annotated, and augmented with precise specifications for the execution of the experiment: for instance, pipetting
prescriptions in microliters, specifications of the times during which the rack
with the vials of the experiment (four to begin with, plus two controls) have
to be kept in a water-bath with a temperature of 37°C, and where probes have
to be divided or aliquots taken. On the one hand, this semi-manufactured
form of the experiment, "Halbzeug" in Hans Blumenberg's sense,[17] gives laboratory activity, particularly that of bigger groups, the coherence and comparability necessary for connected research activities. On the other hand, it
also allows for the generative enactment of a differential reproduction that is
characteristic of the micro-performative process of the empirical generation
of knowledge. For it is only when a group of experimenters follows such half-
codified conventions that deviations in experiments can achieve significance.

It is only in this way that experimental blockages can be stated and eventually resolved, that iterations can be rendered fertile or turned around, that ramifications can be tracked down and peripheral noises excluded or amplified and developed into signals. Achieving this requires such fixtures; they represent, on paper, the ground constellation of experimental systems with their technical apparatus and epistemic components.

Literal technologies of this kind go along with a laboratory identity that can be described as idiosyncratic. The group identifies with these procedural formalisms. Its members pass on elements of a laboratory practice that have proved successful under particular local conditions and yet cannot simply be generalized. They are not made for that purpose, for they are not directed toward the scientific community of whole experimental cultures. Rather, their function is to hold together the activities of a laboratory, to make them easy to grasp for all the members, and—last but not least—to introduce newcomers to the laboratory practices.

A remark on laboratory manuals is appropriate here. In contrast to the procedural protocols, they are not located at the level of particular experimental systems but are directed instead toward the community of entire experimental cultures. A good example is *Molecular Cloning: A Laboratory Manual*, a laboratory handbook by Joseph Sambrook, Edward Fritsch, and Tom Maniatis which appeared in 1982 as a ring binder published by Cold Spring Harbor Laboratory Press.[18] It was intended as a protocol handout for publicizing the molecular techniques of recombining DNA derived from different organisms and their amplification by molecular cloning in bacteria. These methods had developed in the previous decade in biological laboratories worldwide. Since then, generations of molecular biologists have learned the techniques related to molecular cloning with the aid of this laboratory manual. A parade of new editions have adapted it to experimental developments and it is still in use today as *Maniatis*, although its authorship has long since changed. This remarkable history of a laboratory practice manual has only recently begun to receive attention from historians of science.[19] It differs from the regular textbook literature in the natural sciences by not following a theory-guided system representing a specialty under review, but being oriented toward the manipulative practices—and in particular their pitfalls— that are connected with the recombination, and more generally, the alteration of nucleic acids in the test tube as well as their proliferation in experimentally appropriate amounts in bacterial bioreactors. It thus transmits laboratory knowledge and is located at the laboratory bench, either complete or as separate sheets.

Taken together, with regard to such scriptural technologies that are iteratively rewritten again and again, the laboratory can be seen as a scientific writing collective. We are dealing with more than the mere fact that a group of people works conjointly to achieve certain results. In terms of written output, laboratories preserve a certain tradition, an identifiable experimental style that is repeatedly reworked and updated through continuous personnel change—whether in the mode of competition or cooperation. It is precisely this function of the laboratory that is still underappreciated in the long-running discussion about research traditions, or research schools, in the history and sociology of science.[20] Here, social characteristics such as the charismatic leader, the aura of a local institution, or the peculiarity of an interdisciplinary constellation have always been in the foreground. For a phenomenology of the laboratory situation, however, it seems wise to begin on a lower level, the level of the laboratory reality, and to consider the collective notational techniques that play a major role in the genesis of such traditions and schools.

Research traditions are ultimately a result of the material reproduction of local research cultures. The scriptural peculiarities of the laboratories— recipes, rules of action, log sheets, half-standardized experimental schemata, software devised by the scientists themselves—are constitutive for this process. The economy of these written records is still largely unexplored. For a phenomenology of the experiment, however, it is an indispensable part of its discursive materiality, in its historical as well as its actual dimensions. Such records should certainly be highlighted and elucidated as a unique and specific writing genre.

Protocols

Notes, experiment sheets, and laboratory manuals help to prepare and carry out an experiment and to present it more or less neatly and clearly. We shall now turn our attention to forms of writing used at or near the laboratory bench, that result from and are done following an experiment. These sorts of notes are usually called protocols in a narrower and more constrained sense. They record the results of experiments. They are, first of all, preliminary contextualizations of the data resulting from the repertoire of an experiment's traces. They provide a written record in various, more or less regulated manners. Mixed modes of traces and data can also occur, if for instance a paper chromatogram or a dried sequence gel is pasted onto a protocol sheet. For the most part, however, the data take the shape of numbers—that is, they are represented as quantifications or assume related, separate forms of representation such as those described in chapter 1.

The realm in which protocols play their role, adjacent to the experiment, between the bench and an eventual research publication, does not simply represent a preliminary, passive recording of what occurs in the experiment. In fact, it is a realm in which the productive engagement with the matter at hand is continued. Once the experimenter has temporarily retreated from the activity of trace generation, she or he is forced to intervene again here. This is where exploration of the possible arrangements of data begins. What initially appears to be disparate is tentatively translated into patterns. What appears at first to be related is separated into different realms, and what appears to be separate is juxtaposed. Typical written and graphical forms of a condensed representation come into play: lists, tables, curves, and graphs.[21]

As an example, let us focus on figure 5.2. It is an evaluation of the kind of experiments whose general outline is represented by the first figure of this chapter. The protocol material stems from the same laboratory context. We are dealing with experiments that explore the binding behavior of smaller molecules (transfer RNAs) to a poly-molecular complex (ribosomes). Three steps are recognizable in this evaluation process; they can be addressed simultaneously as stages of visualization, as discussed in chapters 2 and 3, on models and on making things visible.

In a first step, such an evaluation takes the shape of lists that start from data strips. In the present case these are counting strips of radioactive probes produced by the instrument used—a liquid scintillation counter—through an overnight counting period.

We should not forget, however, that these columns of counts are already preceded by a series of processing steps which the probes resulting from the experiment have undergone beforehand, and which still belong to the realm of a material processing of the probes. In a first step, the radioactive samples have to be poured over an appropriate filter on which the binding complexes are retained and the unbound radioactivity is washed out. In order to separate the two different kinds of isotopes used in the experiment—radioactive hydrogen and radioactive carbon—the filter disks are then burned in a special apparatus. The products of the burning process, radioactive water and radioactive carbon dioxide, are subsequently separated from each other so that both isotopes can be counted separately. These data are then averaged, subjected to background correction, and converted into molecular masses. As a first result of these processing steps, we have nothing but an array of binding values.

PARAMETER GROUP: 5
ID: UNIVERSAL CPM LONG

01A PROGRAM MODE   1 ->
01B COUNT MODE: FIXED WINDOW: 5- 320
                SQP(I) SINGLE LABEL
                P-32
02 LISTING   Y ->
03 TIME   180 ->
04 COUNTS 1   900000 ->
07 NUMBER OF WINDOWS   3 ->
08 WINDOW 1   5- 320 ->
09 WINDOW 2   120- 650 ->
10A WINDOW 3   5- 960 ->
14 PRINT 1.2.11.21.43.32.39

                    CPM1   CPM2

Oxidizer-   14.41   31.16
Kontrolle   17.42   31.49
              127732.36   67550.27
                   19.74   35.41
                  429.14   238.23
                13523.19   14593.20
               122232.47   64686.06
                 9828.00   10878.59
                  345.78   198.74
                   16.40   32.07

                                        Ø cpm/µl pmol/µl cpm/pmol

2 x 2 µl   CPM1   CPM2
[14C]tRNA 1:10
                  179.41   95.32
                 2040.70   1930.71   996 (-bg 980) 18   55 (54)
                  133.55   80.58
                 1890.71   2057.64
2 x 2 µl   45668.59   26480.89
[14C]Pue tRNA   38.15   58.82
A:10   48685.24   25325.86   24538   3.5   7010
                   22.76   32.41
                 35821.89   14410.17
                 1931.50   1842.60
                                            982 (966) 18   54 (54)
DL
                    CPM1   CPM2

                48837.02   25764.18
                 1854.40   2087.93
                  326.00   90.79
                   13.38   32.41

         tRNA   CPM1   cpm-bg   x 10/5   pmol   % Asik
                17278.15
15   1   23451.35   6173   8848   1.1
                45925.79   28647   40924   5.2   79
                19939.83
15   2   20734.20   795   1135   0.1
                43654.82   23715   33878   4.3   98
                14999.55
6 Kp   1   16633.61   1634   2334   0.3
                34006.68   19607   27152   3.4   91
                15883.08
6 Kp   2   16312.06   425   612   0.08
                26757.49   10874   15534   1.96   96
2 x 2 µl Ac[14C]Phe 55425.80   Ø cpm/µl pmol/µl cpm/pmol
tRNA, A:10   55586.62   27752   3.5   7929

Lists of this kind can be condensed into tables. Tables are synopses of lists. Figure 5.3 shows such a table containing a comparison of several similar experiments. The variable on which the table focuses is the concentration of the divalent magnesium ions in the buffer in which the experiments were conducted. It has a decisive influence on the binding of transfer RNA to the ribosomes and is therefore investigated in extensive series concerning the reactions involved in protein synthesis.

Another form of condensation (and with that, visualization) of data that goes yet a step further are graphs such as those in figure 5.4. What appears juxtaposed in a formless numerical manner in lists and tables assumes a recognizable shape here. For example, one can now make comparisons between experimental runs by introducing a second variable. Both resulting curves are dependent on the magnesium concentration. In addition, they show the reaction of the antibiotic puromycin (PM) to be dependent on the presence of a protein factor—the so-called elongation factor G, written here as EF-G—with the amino acid of the transfer RNA coupled to the ribosome. Curves of this kind have been, perhaps since the middle of the nineteenth century, the most important form of orientation accompanying physiological research. As already mentioned, Emil Du Bois-Reymond lauded the curve early on as a tool for physiological investigations, in his preamble to the *Investigations on Animal Electricity* of 1848. He explicitly valued it as being superior to what he called the "annoying formulas" (*Formelwesen*), particularly in the case of research processes bordering on the unknown.[22] At the same time, he refused to cast the "procedure" as a "command" or prescription: "As is well known, there is nothing more difficult, and at the same time mostly useless than to abstract in this way the rules of a convoluted activity that is neither learned in one stroke nor carried out with consciousness."[23]

Figure 5.5 shows how the shapes of curves can be translated into iconic representations that give an eidetic impression of the kind of molecular movements one expects in such experiments. They can be categorized as visual forms of schematization. In this additional step, the experimental data are condensed into a schematic image that shows the succession of movements of a transfer RNA molecule across the ribosome, through which the particle changes its conformation accordingly. They are symbolized by round and angular contours, as well as by notches. This way of making a molecular movement visible simultaneously leads from a quantitative representation—in the form of lists, tables, and curves—to a qualitative representation that exhibits, in this case, a schematic, iconic character. The data are transformed into an image of successive molecular conformations of the ribosome in the course of protein synthesis. We discussed its historical development in chapter 3.

Handwritten worksheet: "Vergleich der Kurven Exp's 419 | 422 | 423"

T [14C] tRNA

Mg^{++}	V=1 Exp422	(423)	Φ	V=2 (419)	(422)	(423)	Φ	Φ 419+423
4	27	31	29	26	20	26	24	26
6	53	55	54	48	39	50	46	49
8	75	70	72	66	58	68	64	67
10	94	79	86	87	74	82	82	84.5
12	99	94	96	91	82	92	92	91.5
15	100	100	100	100	100	100	100	100
6 Sp.Gpd	113	107	110	95	89	110	98	102.5
V 100% =	0.83	0.98	0.91	1.94		1.68		1.81

P site : 418 + 420
A site (1x) : 424
A site (2x) : 419 + 425

Mg^{++}	P site 418	420	Φ	A site (1x) 424	A site (2x) 419	425	Φ	hom
4	20	24	22	22	20	18	19	19
6	42	41	41.5	36	40	30	35	36
8	73	74	73.5	64	74	55	63	64
10	89	96	92.5	82	91	68	79.5	81
12	100	99	99.5	91	100	87	93.5	95
15	100	100	100	100	97	100	98.5	100
6 Sp/Gpd	113	110	111.5	97	98	123	110.5	112
100% =	0.9	0.9		0.55	(0.35)	0.56		

FIGURE 5.3. Comparison of similar experiments (nos. 419, 422, 423) in the form of a table
Source: Archive of the author.

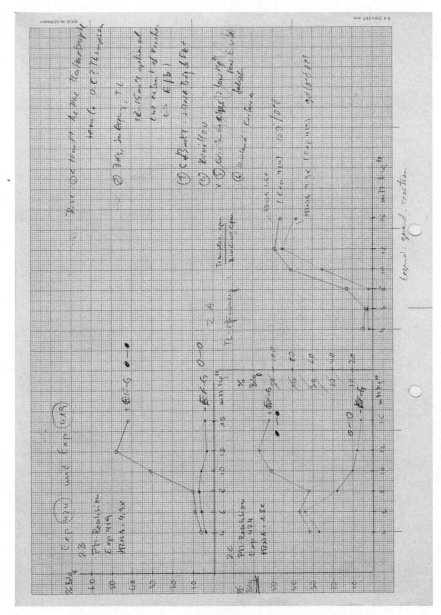

FIGURE 5.4. Experiments nos. 419 and 424, evaluation in the form of a curve

Source: Archive of the author.

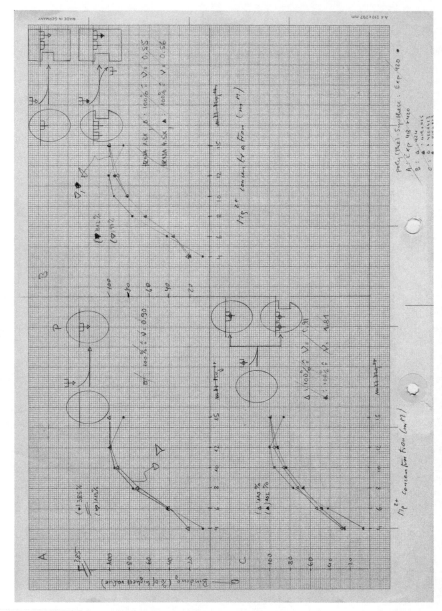

FIGURE 5.5. Experiments nos. 418 to 425, evaluation in the form of curves and icons

Source: Archive of the author.

This paperwork is a rough draft, often scribbled, at times with crossings-out, revisions, and annotations, and sometimes bearing the traces of collective reflection and processing. It does not aspire to the sphere of a printed communication addressing the scientific community outside the laboratory; instead, it remains on the side of the tentative practices of the experimental activity. Such notes still belong to the knowledge regime of the laboratory and are not made for wider circulation.

One of the decisive functions of writing up, and sometimes still literally chalking down at the laboratory bench, is to reduce the dimensions of the experimental arrangement. In its trivial sense, this means that the spatio-temporal arrangement of the experiment is basically stretched across a two-dimensional frame. This makes the laboratory easier to handle and more transportable.[24] What is lost in terms of dimensionality is gained in terms of portability. The reduction to the plane simultaneously requires and furthers the exploration of new possibilities of arrangement. Successive events can be represented simultaneously, simultaneous processes can be disentangled, and temporal relations can be recorded as spatial relations. Analogous to the experiment itself, the procedures being applied in laboratory protocols produce effects of either dilatation or compression. As we can see from the examples given here, the latter can pass through different stages in which, according to the ordering and compression modes of data, new patterns become visible that impact on the future course of experimentation and can turn out to be either insignificant or phenomenally relevant.

In a well-conducted laboratory protocol, it is essential that all these deductions and reductions remain reversible. A protocol is epistemically productive if the chain of transformations can be followed back and at certain points resolved and re-nested.[25] As a rule, laboratory protocols represent the memory of whole series or strata of experiments; it is primarily through these that individual data are imbued with contextual significance. Accordingly, the transformation of the laboratory into a writing surface, the conversion of the memory space from a mere chronology into a meaningful sign patchwork of indices, icons, and symbols, involves more functions than merely condensing and dilating. Laboratory notes and laboratory protocols provide new resources and materials that infuse research with their characteristic contours and prevent its premature closure. As a space of representation, they have an eminently productive character that is indispensable for experimentation. In addition, they structure the memory space that accompanies the trajectories of experimental systems.

In the realm of such scientific writing, which is not yet directed toward and guided by a definitely imagined product, individual styles can find expression in the creation of scientific novelty. In contrast to the largely standardized modes of writing in scientific publications that exclude such idiosyncrasies, the protocol becomes a space of "new combinations,"[26] in which the tamed phantasy of research can unfold and act out its creative imaginations.

Annotations to research are residues on paper of exploratory movements at the workbench. We find ourselves here in the space of the pre-normative, in which the opportunism of the process of knowledge acquisition can flourish unhindered, in the space of the trial in the deeper sense of the word that is constitutive for natural scientific research[27]—as well as for cultural investigation.[28] A trial in this sense is not a test, not a check whose aim is always the detection of a greater or lesser quantity, or a yes or no for borderline cases. It is, rather, a movement of exploration, a play with possible positionings, an open arrangement, a space of contingency. In a research process, such exploratory procedures transpose the unfathomed explanatory potential of experimental systems into a combination of graphic images. They unfold it in a space of representation that is not restricted by rigid boundaries of compatibility and exclusion criteria, a space that is not yet bound to the logic of communicating more or less definitive results.

Repositories

Are there collective equivalents to these individual scientific forms and moves of preliminary writing? Are there written forms that preserve traces that, on the one hand, no longer accompany a singular experiment or a series of experiments, but on the other hand, do not consist of definitive statements and therefore still remain in the space of epistemic things whose investigation they accompany and comment on? And what could possibly be the characteristics of this type of notes that, along with the others, we could address, with Michel Foucault, as the "discourse-objects"—*discours-objets*—of a laboratory archaeology?[29]

In the interstices between the laboratory bench and the professionally organized discourse of scientific communities with their congress posters, journal articles, reviews, and textbooks, one finds different categories of collective writing and of keeping data. We shall briefly rehearse them in concluding this chapter. This function has traditionally been filled by lists, tables, and other forms of scientific bookkeeping in the broadest sense.[30] For a history of science dating further back, meteorological records collected by lay meteorologists can be mentioned here.[31] Astronomical data catalogs kept by observatories,[32] and the tablework of hospitals and asylums also contributed to these records.[33] In the research undertakings to which they are connected, they function, first, as registers or archives—that is, systems of keeping data at hand that supply collections of bits and pieces of information which aid the design of new experiments or serve as background for further observations. Second, they take on the role of permanently actualized primary data banks by continually registering new research results or observations, thus making them accessible for the collective work of whole networks of laboratories, clinics, or other health institutions.

Today, these extended functions of the counting techniques of laboratories and observatories are fulfilled mainly by forms of electronic storage, representation, retrieval, and transmission of data in data banks. A prominent example from the realm of the life sciences are the sequence data banks of nucleic acids and proteins[34] on which molecular geneticists and gene technologists rely in constructing probes and comparing results, and into which they eventually feed their own sequencing results. Such digital information pools represent a peculiar form of networking with respect to the research process. However, they also become the locus of data experiments of sorts that, with the help of algorithmic tools, engender new research questions that only become possible with such massive databases.[35] However, there seems to be no doubt that in the foreseeable future the tentative writing and drawing by hand in the data space will remain a "procedure of research,"[36] just as sketching with a pencil has not been ousted by computer-assisted drawing in architecture.[37]

To sum up: in its material constitution, the laboratory is a locus of experimental trajectories with their phenomenal expressions as described in chapters 1 to 4. This chapter has shown that just as importantly, to quote Friedrich Kittler, it is a "writing system," an *Aufschreibesystem*,[38] an indispensable space of notation for emergent knowledge.

PART II

Supra-Experimentality

Shapes of Time

This chapter will consider the forms that the temporal course of epistemic processes can assume. At stake is thus a phenomenology of time, not time as an experimental object.[1] In her essay on time, Helga Nowotny complained that the "discursive exchange about time is underdeveloped,"[2] and indeed, since the pioneering works of Hayden White,[3] Krzysztof Pomian,[4] and Reinhart Kosellek,[5] the field of reflection over the temporal shapes of the historical has largely lain fallow. In this chapter we are less concerned with forms of narration—there will be more on them in chapter 8—than with the flow patterns of material systems. The chapter begins by recalling the art historian George Kubler's attempt to trace the shapes of time in the history of the arts and to connect them at the same time with the history of the sciences. In the second part I shall sketch a number of configurations of time related to experimental systems, epistemic things, and cultures of experimentation.

George Kubler: *The Shape of Time*

Two things particularly fascinated me when I first encountered George Kubler's *The Shape of Time: Remarks on the History of Things* three decades ago.[6] The first was the scansions of temporality in which artistic as well as scientific developmental processes take place, or better, *through* which these processes take shape. The second was the vehemence with which Kubler defended the role of *things* in these developmental processes: in other words, the materiality of the aesthetic and epistemic objects around which these processes revolve, and their role as genuine actors in the constitution of the respective shapes of time. The two themes are indeed inseparably related, as indicated by the title and subtitle of the book.

The art historian George Kubler was (like his American contemporary, the historian of science Thomas Kuhn) a thinker in terms of structures. Though the parallels should not be exaggerated and the differences should not be overlooked, these two thinkers can be aligned with two authors that simultaneously left their traces in continental European thought: Michel Foucault and Jacques Derrida. Both have also served as points of reference on several occasions in the present book. They mark a conjuncture whose derivations have to this day not been fully understood.[7] Foucault was more interested in the contours of an *episteme*, that is, of spaces of knowledge, and in this respect can be compared with Kuhn and his interest in the structures of scientific paradigms. Derrida, for his part, was more concerned with the diachronic flow structures of the historical process and its conceptualization, which allies him with Kubler in the latter's interest in the shapes of time to be found in the artistic realm. We can thus distinguish thinkers of more vertical from those of more horizontal concatenations.

Kubler did not see the flow of time in the field of the arts as a massive, homogenous stream with phases of rising and ebbing, cesurae and cascades; rather, he viewed it as a structure composed of elementary units, each carrying its own temporality. One could also describe these units as contingency networks: "We can imagine the flow of time as assuming the shapes of fibrous bundles. . . . The cultural bundles therefore consist of variegated fibrous lengths of happening, mostly long, and many brief. They are juxtaposed largely by chance and rarely by conscious forethought or rigorous planning."[8] The image of the fibrous bundle brings together both structural dimensions and forms of concatenation, which are manifested in the transverse as well as the longitudinal section of the fibers.

For the characterization of historical times, the general historiography as well as the historiography of the arts and the sciences have repeatedly referred to the *longue durée* of biological evolution or the *courte durée* of individual human development with their specific timespans. Kubler decidedly distances himself from both of them. His aim is to capture genuinely cultural-historical processes between these extremes in their movement patterns—in short, the shapes of cultural time sui generis between ontogenesis and phylogenesis, between biographical trajectories and evolution.[9] In doing so, Kubler particularly targets the peculiar characters and properties of the *objects* of culture, the things with which the arts are concerned.[10] Above all, unlike so many cultural and art historians before him, he is not interested in defining successive or repeating stages, or even laws of cultural development. His goal is rather to provide a conceptual arsenal that makes it possible to capture temporal patterns or trajectories of sequences of objects. His interest is directed toward identifying patterns of "moderate compressibility" (to borrow an expression from complexity theory)[11]—in other words, figures of temporal condensation of a medium range that he calls "shapes."

The aspiration to reflect the categories in which the history of works of art can be represented aims not only at capturing meaningful temporal figurations in their formal and symbolic content, but at arriving at something like a material semantics of things. Ernst Cassirer formulated the same aspiration in his late Gothenburg studies on *The Logic of the Cultural Sciences*: "What emerges is a 'meaning,' which is not absorbed by what is merely physical, but is embodied upon and within it; it is the factor common to all that content which we designate as 'culture.'"[12] In view of this formulation, it may seem like a little irony of history that Kubler, in the very first sentence of his treatise, set himself the task of overcoming "Cassirer's partial definition of art as symbolic language" that "has dominated art studies in our century."[13] Obviously, Kubler was familiar only with the Cassirer of the *Philosophy of Symbolic Forms* when he wrote his essay in the early 1960s.[14] He may have encountered the three volumes of this monumental work during his studies in Berlin and Munich in the early 1930s, or else during Cassirer's period of exile at Yale in the early 1940s. That question can remain unresolved at this point.

Kubler's short essay sketches a history of art that goes way beyond a concentration on the symbolic content of works of art. What is more, he also pleads—if only programmatically—for a rapprochement of the history of the arts and sciences under a common denominator: "Although both the history of art and the history of science have the same recent origins in the eighteenth-century learning of the European Enlightenment, our inherited habit of separating art from science goes back to the ancient division between liberal and mechanical arts. The separation has had most regrettable consequences. A principal one is our long reluctance to view the processes common to both art and science in the same historical perspective."[15]

In stating their commonalities, Kubler had no intention of blurring the differences between artistic and scientific things, and certainly not those between works of art and technical things. On the contrary, he stressed their distinctness: "Although a common gradient connects use and beauty, the two are irreducibly different: no tool can be fully explained as a work of art, nor vice versa."[16] However, he saw comparability in the processes of the genesis of epistemic things (including the technical things emerging from them), and of works of art, in what he called their "common traits of invention, change, and obsolescence":[17] in the temporal dynamics of their becoming, and consequently in the structures of their emergence.

This induced Kubler to take a radical turn. To this day, both the history of art and the history of science are replete with reports about great men in the mode of what Edgar Zilsel once aptly called the "genius religion."[18] Novelty is represented in this context not only as the result of the genius-like intuitions of an extraordinary individual, but usually also as a brainchild: as the sudden inspiration of an exceptionally gifted brilliance. Kubler is vehemently opposed to this idea. In his book, novelty is presented not as the result of the distinguished illumination of a genius, but as the unprecedented result of a retrojection. He compares the situation of the artist with that of a miner digging for ore who stands at the end of a shaft and can assess the tunnels that others before him have dug, but has no guarantee that the direction he is about to take will not lead him astray.[19] Kubler's artists—and scientists—thus move in a terrain the materiality of which they inexorably confront. Instead of their dreams, they are scarred by the paths that have been trodden before them and that reveal their future options only if they are ready to take the risk of a decision and to confront the challenges of the material at the face. That by no means implies denying the uniqueness of the art works: "Works of art are as unique and irreplaceable as tools are common and expendable,"[20] Kubler stresses, in a comparison between useful things and things that are not predominantly destined to be used and consumed. However (and this is the decisive point), he sees the arts—like Kuhn sees the sciences—as "a process that move[s] steadily *from* primitive beginnings but *toward* no goal."[21] We are therefore concerned not with an "evolution-toward-what-we-wish-to-know," but with an "evolution-from-what-we-know"[22]—a development whose course is only determined by seizing the moment under the given conditions of possibility. Consequently, it is always also the result of the pathways that have already been taken, but from which we can nevertheless not deduce whether and where we will hit the next bonanza, or whether and where the next escarpment will be.

This is exactly where we find the decisive point of Kubler's vision of a rapprochement between the history of science and the history of art. And it is here that Kubler brings into play the concept of "series" or "sequence" with its variants. He distinguishes between open and closed sequences; intermittent and arrested classes; extended, wandering, and simultaneous series of objects of art, each characterized by its peculiar forms of manifestation.[23] To retain Kubler's imagery, such series or sequences of art works—or formal approaches to problems, for that matter—are to be considered as temporally stretched individual fibers in the fibrous bundle of art. Kubler's own attempt at a definition of the concept of sequence reads as follows: "The closest definition of a formal sequence that we now can venture is to affirm it as a historical network of gradually altered repetitions of the same trait. The sequence might therefore be described as having an armature. In cross section let us say that it shows a network, a mesh, or a cluster of subordinate traits; and in long section that it has a fiber-like structure of temporal stages, all recognizably similar, yet altering in their mesh from beginning to end."[24]

Sequences thus form ensembles. They can proliferate, they can exhaust themselves, they can ramify and establish new series, they can merge or become intertwined, transposed temporally and spatially—there is a whole arsenal of forms of movement connected to them. And it is always their material characteristics that underlie the movement patterns that are actually executed. As we shall see, such temporal forms of movement are also characteristic for experimental systems.

Epistemic Trajectories

In the second part of this chapter, I would like to situate Kubler's views on "series," "sequences," and "solution chains" in the wealth of forms of art works in relation to my own deliberations on epistemic things, their technical coatings, and their trajectories in the history of the sciences. No strict equivalences should be expected; but we should expect instead some family resemblances in the temporal structures. Before doing this, however, I would like to recall the relevant passages in the work of Ludwik Fleck that I mentioned at the beginning of chapter 2. Had Kubler known Fleck's work, he could have referred to him as a source from the history and epistemology of the sciences. In *Genesis and Development of a Scientific Fact* Fleck states that "It is very difficult, if not impossible, to give an accurate historical account of a scientific discipline. Many developing strands of thought intersect and interact with one another. All of these would have to be represented, first, as continuous lines of development and, second, in every one of their mutual connections. Third, the main direction of the development, taken as an idealized average, would have to be drawn separately and at the same time."[25] Elsewhere in his book, which only belatedly became widely known and read, Fleck emphasizes that single experiments only gain significance in series or "systems."[26] At the same time, however, that means that in the course of a "line of development," as defined by Fleck, the experiments also tend to be *"carried along by a system of earlier experiments and decisions."*[27]

Kubler's imagery of tunnels and shafts returns here, but straightened, in the metaphor of a tow rope. In this form, however, the image tends to disguise openness toward the future rather than giving it an adequate expression. It is obvious that Fleck was particularly interested in kinds of closure and in the

accompanying "harmony of illusions."[28] Kubler also described such series—
"closed sequences" and "arrested classes"[29]—and in the sciences, too, experimental series and systems can eventually lead to dead ends, a permanent risk
for everybody working in a laboratory context.

Let me now briefly present five kinds of trajectories or sequences that can
be specified in the empirical sciences and that constitute an equal number of
shapes of time with their own traits and characteristics.

TRAJECTORIES OF EXPERIMENTAL SYSTEMS

First, there are the experimental systems I discussed in chapter 1 as the frames
for the production of experimental traces and data. As a rule, they form trajectories that can stretch over longer periods of time.[30] Their lifespan often
coincides with the productivity cycle of a particular research group. Experimental systems are loose articulations of materials, research technologies,
and procedures, whose couplings are, however, sufficient to shape a research
process—in the framework of a certain stage of the development of a research
field and around a particular research object—in such a way that there is hope
of discovering something new about the target object. Although experimental
systems are temporally limited and localized research environments, they are
nevertheless endowed with an underlying material continuity within a limited
time span. Along this axis of continuity, they change permanently, albeit in a
differential manner that operates on displacements rather than disruptions.

However, it is always possible that such systems suddenly end, particularly if an alternative system promises to advance the same problem complex
more efficiently. The history of the sciences is full of such abandoned systems.
To give one example: in the late 1950s, three different experimental systems
were positioned in the ambit of a problem that then appeared imminent:
the elucidation of the genetic code. Sydney Brenner and Francis Crick's system was centered around the analysis of the nucleic acid of bacteriophages;
Heinz-Günter Wittmann's system concentrated on the analysis of mutants of
the tobacco mosaic virus (TMV); and the system used by Heinrich Matthaei
and Marshall Nirenberg aimed at an analysis of bacterial protein biosynthesis. Although the latter had not been established from the outset with the
genetic code in mind, in the early 1960s it rendered the two other systems obsolete.[31] This was essentially due to the temporal characteristic of the in vitro
protein synthesis system. It allowed for carrying out a decisively denser succession of experiments and a much more direct interpretation of the results.
This effected a temporal condensation of the experimental work comprising
several orders of magnitude. The time expenditure, for instance, of infecting

tobacco plants with TMV mutants and the subsequent tedious analysis of their coat protein sequences was completely out of proportion to the isolation of the components of the protein synthesis apparatus from bacterial cells and the laboratory synthesis of the oligonucleotides necessary for the experiments.

MIGRATIONS OF EPISTEMIC THINGS

This example plainly shows that experimental system and epistemic target have to be distinguished from one another. The epistemic objects are the materially and theoretically underdetermined entities on which the scientific interest of those who run an experimental system focuses. The latter, in turn, consists of the research technologies that help to give contours to the things of epistemic interest. In a concrete research environment, the investigated phenomenon and the research technology are specifically correlated—they even condition each other. And yet, materially they can be of a very different nature: for instance, an enzyme and an ultracentrifuge. As expounded in chapter 4, there will be no system effects without an epistemically productive interface between phenomenon and technology. An instrument that cannot be adjusted to an epistemic object has no scientific value; and an epistemic object that cannot be made compliant with the instrument in one way or another is not a phenomenon that can be investigated under scientific auspices.

The concept of the epistemic has a double connotation here. First, it designates the object of knowledge in its essentially underdetermined form, a circumstance that makes it relevant for research in the first place. That is precisely why I also speak about epistemic things. The notion of "thing" carries with it something of the indeterminacy that lies at the core of research objects. That is due not least to the etymology of the word that refers to the circumstance that we are concerned with a thing that is put on trial, brought before court. The concept of a technical object, in contrast, is normally associated with fixed and sharply delineated contours. It is not up for debate; it has a closed character. Second, the concept of the epistemic, in contrast to the technical, refers to the circumstance that epistemic things only take shape and exist in an instrumentally mediated form, and are therefore ontologically transitory in form and duration. The epistemic and the technical consequently remain mutually dependent and draw their respective specificity from that relation.

As already mentioned in passing, the material continuity of an experimental system does not imply the continuity of the epistemic things dealt with in it. On the contrary, at certain turning points in the development of an experimental system abrupt changes in the realm of the epistemic can happen without necessarily having a disruptive effect on the system. For example: in 1894 Carl Correns began to cross corn varieties; shortly thereafter he branched out into pea varieties, in the hope of producing the biological phenomenon of xenia—the direct influence of the pollen on the mother plant—in an experimentally clear fashion, and then being able to investigate its physiology more closely.[32] In this he was successful with his maize varieties, although not with his peas, where he was struck instead by the regularity with which, in the second experimental generation, the characters of that parent part that had vanished in the first crossing generation reappeared. Following this new track required only comparatively small modifications, mainly concerning the amplification of certain hybridizations. On the whole Correns could continue his experimental regime and achieve robust statistical statements about the progeny ratios of his crosses. Even the marginal observations made on this matter when the xenia aspect was still prevalent in the earlier experiments could be re-evaluated and reintegrated into the later ones. Consequently, the search for traits of the paternal plant in the seeds and fruits of the mother plant was replaced by the rediscovery of Gregor Mendel's rules, resulting in Mendel's experimental and theoretical corollaries moving now to center stage. The new epistemic things were the "Anlagen"—to use Correns's terminology—that were apparently inherited independently of each other. They appeared to be responsible for the specific traits of the plants. Under the label of "genes," they would soon begin their trajectory through the biological experimental systems of genetics that would extend through the twentieth century.[33]

By hinting at this extended trajectory, we are touching on the second point to be made here. The trajectory of epistemic things is not identical with the trajectory of experimental systems, not only because the former can be displaced from an existing system and replaced by a different epistemic thing, but also because they can jump from one experimental system to another. As a rule, they emerge from a particular experimental system and are developed there to a certain degree. That an epistemic object remains completely within the context of a single experimental system happens rather infrequently.

Staying with the example of the genes just mentioned—one of the cardinal epistemic things in the life sciences of the twentieth century—experimental hybridization systems dominated the first four decades of the century, supplemented by the cytological preparation of giant chromosomes. The emergence of molecular biology led to the addition of biophysical systems such as X-ray structure analysis of biological macromolecules and the mutagenesis of viruses induced by X-rays, as well as biochemical in vitro systems like the above-mentioned protein biosynthesis, to recall just a few of the many experimental systems that contributed to analyzing and characterizing genes as things of epistemic interest. The emergence of gene technology from the 1970s onward led to the manipulation of genes in living cells by means of the molecular techniques of molecular biology—techniques that in turn resulted from the transformation of molecular epistemic things into molecular technical tools.

All that cannot and shall not be expanded upon in detail here.[34] What I would like to take from this discussion is that epistemic things can migrate from one experimental context to another. These migrations are not linear and directed, but ramified, multiple, and intermeshing. On the one hand they promote the stabilization of certain aspects of the epistemic things in question, which means that they can themselves eventually assume an instrumental character.[35] Thus, in contexts of biotechnological production, precision-tailored genes function today as molecular replication machines. On the other hand, they do not come to rest as epistemic things, because the multiplication of their uses as tools is accompanied by the emergence of new blind spots. Through migrations of this kind, the experimental systems involved become interconnected to a certain extent, resulting in multiply networked experimental webs that are mutually associated through materially mediated pathways. In this way, and in the context of experimental cultures, experimental systems that are only marginally related because they differ in terms of their relevant technical components can nevertheless communicate with each other.

THE PATHWAYS OF TECHNICAL OBJECTS

The technical components of experimental systems—that is, the research technologies and other standardized parts of such systems—exhibit characteristic migratory movements and trajectories as well. In contrast to the history and migratory movements of experimental systems and epistemic things, there is an extensive literature on the history and dissemination of research technologies, so there is no need to focus on this point in detail.[36] What interests us here are the temporal shapes, the movements in time that research technologies perform in the space of the epistemic. I showed in chapter 4 that the action radius of experimental systems can be widened and altered by grafting new research technologies onto them. In the phase of their emergence, new research technologies mostly crystallize by handling and manipulating the materials, procedures, and energy forms that determine their core structure, as has been shown repeatedly and in an exemplary fashion in the case of microscopy.[37] Here it is often previously unimagined effects of the investigated material that open perspectives on a new technology—such as the use of accelerated electrons for the production of enlarged representations of material structures, in the case of electron microscopy. In the ensuing development, instrument and object of investigation enter into a tight reciprocal interaction with each other. This becomes particularly evident when looking at the uses of electron microscopy for the enlarged representation of organic materials. Epistemic object and research technology enter into a process of mutual instruction that can trigger epistemic as well as technical effects. It is not unusual, therefore, that research technologies themselves—again, electron microscopy is a good example—will move over longer timespans in an experimental, essentially epistemic realm and can even assume the status of something like a technical experimental system sui generis. Sooner or later, however, most research technologies become encased in more or less closed packages. For their users, that transforms them into black boxes.[38] This prevents them from being tinkered with in the laboratory; only the intersections with the system in question remain open for the experimental grasp. However, their horizon of application is widened. Their migration is facilitated, and they can be integrated into a greater number of experimental environments.

Something akin to consignment to black boxes can also happen to experimental systems as a whole: they can be transformed into test systems and become part of other experimental setups as subroutines. This shows that although all these components and configurations of the experimental process can and must be differentiated, we are still by no means dealing with elements that are locked into these roles forever. Rather, they can change places in a dynamic manner. Only concrete historical analysis can convincingly show what function a particular element fulfills in a given experimental context.

ON THE MOVEMENT PATTERNS OF
EXPERIMENTAL CULTURES

The migration phenomena just pointed at, whether related to epistemic things or to research technologies, decisively contribute to the constitution of experimental cultures; they make them appear as multiply reticulated ensembles of experimental trajectories of the different kinds of components described above. Their main characteristics will be expounded in more detail in chapter 7, using the example of the biological in vitro cultures of the twentieth century. Therefore, I can be brief here. In compound experimental structures as described in more detail in chapter 4, fusions or hybridizations of experimental systems that originally developed separately can happen. It can also happen that having become too rich in options, experimental systems split up. Within an experimental culture, such bifurcations or even ramifications can lead to niches in which the exchange between the systems involved is particularly intense. Here again, the floating character of any classification becomes plain to see. The ramification of experimental systems could have fitted equally well in chapter 4, and the hybridization described in that chapter could have been inserted here. Such fluidity stems from the fact that the transition from infra-experimentality to supra-experimentality itself—notwithstanding the significance of the distinction—is a floating one. Whereas this chapter focuses on the temporal shapes inherent in the parts of the experimental web, chapter 4 concentrated on its mechanisms and the respective micrological movement patterns that belong to them.

ORGANISMIC MIGRATIONS

I would like to conclude the broad range of these time patterns with a particular case. It is neither an experimental system nor an experimental culture, neither an epistemic nor an exclusively technical thing. We could perhaps capture it with the concept of "strategic research material," which we owe to Robert Merton.[39] Merton uses it to designate materials that for certain reasons are particularly suitable for the study of natural phenomena that show themselves here in a pregnant form, or can be made accessible in a particularly obvious way. Such materials often have a long history that does not necessarily have to be situated in the space of the epistemic to begin with. They can, however, end up in science on their way through completely different cultural contexts.

In the biological sciences this often concerns particular organisms. In the early twentieth century, expressions such as "material for experimental purposes,"[40] "laboratory object,"[41] "laboratory animal,"[42] or experimental organism were current; today, the concept of "model organism," as dealt with in chapter 3, has gained the upper hand. In a very special way, model organisms are carriers of shapes of time that are essentially determined by their reproductive cycles. In the history of biology in the twentieth century, a tendency can be observed toward the economization of time, and with that, the transition to model organisms that reproduce more quickly. Early classical genetics still worked with plants that were bound by an annual reproduction rhythm. In the case of insects—in particular the fruit fly *Drosophila melanogaster*[43]—model organisms were introduced into the laboratory whose procreation cycles were reduced to months or even weeks. Early molecular biology finally specialized in unicellular organisms such as lower fungi or bacteria with a reproduction cycle of hours or less. The impact of these organism-specific time regimes on the rhythmicity of experimental systems in the life sciences can hardly be overestimated.

The model organism on which I would like to focus here as an example is one with a medium-range reproduction cycle: the flour moth *Ephestia kühniella* (Zeller). It is an unspectacular insect with gray-streaked wings from the family of pyralid moths. Assumed to be native to Central America, it found its way to Europe in the nineteenth century in the wake of the colonial traffic. The precise time of its arrival is unknown. In Europe, it first attracted attention as an unwanted guest of grain mills and the flour processing trade, where under Central European weather conditions it bred in three generations around the year. Consequently, the moth began its scientific career as a pest insect, an object of investigation of applied entomology, and the aim was to control its spread or even to eradicate it.

In its role as a harmful insect the flour moth first fell into the hands of the agricultural scientist and phytopathologist Julius Kühn. Kühn had been conducting research and teaching at the University of Halle (Saale) since 1862, and had founded the university's Agricultural Institute in 1863. In the summer of 1877, he collected several specimens of the moth from a mill that processed American grain and sent them to the entomologist Philipp Christoph Zeller in Stettin for determination. Zeller, who had studied at the University of Berlin, was a specialist in smaller moths and a longtime secretary of the Entomological Society of Stettin. His description of the insect under the scientific name *Ephestia kühniella* (Zeller) appeared in 1879 in the *Stettiner Entomologische Zeitung*. The entomological gazette of Stettin and the society had been founded by the sugar refiner and beetle collector Carl August Dohrn after he returned to his hometown of Stettin from a voyage around the world. The trip on which he had embarked in 1831 led him to South America, among other regions. It took place in parallel to Charles Darwin's far more famous voyage on the *Beagle*. Carl August's son, the founder of the Zoological Station in Naples, would later become a lifelong admirer of Darwin.

The flour moth received increased attention during the First World War. It concentrated in the big mills that became centralized in the course of the war, and propagated there en masse. The parasitologist Albrecht Hase, who specialized in insecticides but also in biological warfare agents, was the first to explore their mass culture as a laboratory insect in the early 1920s at the Biologische Reichsanstalt in Berlin-Dahlem. Above all, he was interested in the biological battle against the flour moth using the ichneumon *Habrobracon juglandis*, which deposits its eggs in the caterpillars of *Ephestia*.

In the 1920s, the Göttingen zoologist Alfred Kühn—not to be confused with Julius Kühn—adopted the mass cultivation procedures Hase had developed on an industrial scale. At that time Alfred Kühn was interested in, among other things, the variation of *Habrobracon* that constituted the dissertation subject of his student Egon Schlottke.[44] Moreover, Kühn and Schlottke used the flour moth *Ephestia* as a feed animal. Presumably by chance, they became aware of conspicuous wing mutations in the moth. As a consequence, and together with his doctoral student Karl Henke, Alfred Kühn embarked on a study of the genetic, physiological, and developmental aspects of the wing coloration and wing pattern of *Ephestia*. But this long-term endeavor, in the course of which, according to their report, the two zoologists inspected more than a hundred thousand moths, failed to bring the success they had expected.

The situation changed when, five years into this endeavor, Kühn's technical assistant Veronika Bartels detected a red-eyed mutant of the flour moth in the mass cultures. It soon became clear that the inheritance of the red eye color followed a recessive Mendelian distribution. The mutant, so it appeared, lacked a substance that obviously circulated freely in the body of the wild type, as Kühn's coworker Ernst Caspari showed in a series of transplantation experiments that were both ingenious and simple.[45] With that, *Ephestia kühniella* had advanced to become the model organism of emerging physiological and developmental genetics, as Kühn called it. It would be another ten years, however, before the substance under observation was characterized biochemically. It turned out that it was not a hormone, as initially assumed, but an intermediate product in the composition of the eye pigment. This substance could not be formed from its educt if, as the result of a gene defect, an enzyme that catalyzed this intermediate step was missing. In this way, all the ingredients of biochemical genetics were identified: gene action chains consisting of genes and gene products (enzymes, or ferments in the language of the group) as well as metabolic chains whose intermediate steps were catalyzed by the enzymes.

The only problem was that the experimental system built up around *Ephestia*, while allowing access to this particular metabolic chain, did not permit an assessment of the genes and their products. The genome of the moth with its thirty or so chromosomes was too complicated to achieve this. Thus, the life of the flour moth as a genetic-physiological model organism remained restricted to a narrow research window between the two world wars. Fungi such as *Neurospora crassa* and yeast, bacteria such as *Escherichia coli*, and viruses such as Tobacco Mosaic Virus, as well as bacteriophages, would soon replace it as the new model organisms of the burgeoning fields of molecular biology and molecular genetics.

The scientific existence of the flour moth thus formed a chain, with its starting point around the middle of the nineteenth century and its end point around the mid-twentieth century. As an object of knowledge, it began its career as an unwanted colonial import that made an impact as a food pest; it advanced to become a taxonomic unit in entomology, gained new life as an experimental organism in biological insect control, and finally mutated into a model organism in the epistemic space of emerging physiological genetics. Then the moth as a model organism disappeared from the scene of knowledge. The French physiologist Claude Bernard once remarked in his notes that the epistemic chains of physiology would constitute a historical connection, but that the links of the chain could by no means be derived from each other by logical necessity.[46]

A similar story can be told about Tobacco Mosaic Virus.[47] It began with the identification of a *contagium vivum fluidum*—that is, an agent that proved to be non-filterable—in the hands of Martinus Willem Beijerinck at the Agricultural College of Wageningen at the end of the nineteenth century. In the 1930s, the virus advanced to become the first biomolecular complex ever crystallized, by the biochemist Wendell Stanley at the Rockefeller Institute for Medical Research in Princeton. From there it continued its career as a strategic biological research material in early electron microscopy before it became a model, in molecular biology, of a nucleic acid in which genetic mutations could be induced and localized. After it had played its part in the elucidation of the genetic code, it returned in the ensuing heyday of molecular biology to the agricultural realm from whence it arose. It has become an instrument of gene technology today, helping to create resistant crops.

A history of science that orients itself toward a history of things will have to follow such migrations, and to strike sparks from such chains and nexuses of contingency. Against such a background, Isabelle Stengers's aforementioned "nomadic concepts," and the migratory entities that Peter Howlett and Mary Morgan captured with the traditional category of "facts," may well appear in a new light.[48]

Epistemic Matters, Artistic Matters

Despite the multiplicity of nexuses that can be observed in scientific experimental cultures over longer periods of time, it remains beyond doubt that the dynamics of the individual systems that interact in experimental cultures is determined, in the last instance, by their proper times.[49] These intrinsic times also determine the points in time at which epistemic effects may happen in particular systems, without any guarantee that such effects will actually materialize. Here again, we can cite Kubler: "Sequence classing stresses the internal coherence of events, all while it shows the sporadic, unpredictable, and irregular nature of their occurrence."[50] And we can concur with his conclusion for the sciences as well: "The field of history contains many circuits which never close. The presence of the conditions for an event does not guarantee the occurrence of that event."[51] Consequently, scientists—as well as artists— have to acquaint themselves with the micrological givens of a particular system and its peculiar temporality in order to position themselves to capture its options from within. The "ingenious ideas" that are talked of in hagiographies consist, as a rule, in researchers' becoming aware of an option such a system offers by displacing one of its elements, inserting a new technique, or pursuing a signal that has surfaced in a hidden corner of the system. This also holds for the dizzying heights of theoretical physics, as no less an expert than Werner Heisenberg reminded us by stressing: "The modern theories did not result from revolutionary ideas that would have been carried into the exact natural sciences from outside; rather, they have been forced upon research in its endeavors to consequently complete the program of classical physics. . . . It goes without saying that everywhere experimental research is the necessary precondition for theoretical insights, and that fundamental progress can be achieved only under the pressure of experimental results, not through speculations."[52] Or as the molecular biologist Mahlon Hoagland once put it: "[In] science a new vision of reality arises from mystery by the action of experimentation." And elsewhere: "In science, an idea can become substance only if it fits into a dynamic accumulating body of knowledge."[53]

In order to be able to do so, however, researchers must develop a particular affinity for the system in and with which they work. So too, Kubler asserts with respect to the artist, "Each sequence affords the opportunities of its particular systematic age to only that group having the temperamental conditions for a good entrance."[54] Put somewhat more colloquially, in scientific circles one repeatedly hears the phrase about being "in the right place at the right time." It implicitly refers to two points that coincide with what has been said about the epistemic space: the fine, multiply articulated structure of an experimental culture and the specific developmental stage, or the age of an experimental system. And it refers to the circumstance that in the interaction of a scientist with his or her system, it is the latter that prevails.

Of course, no strict parallel can be expected in the sense of a one-to-one homology between the exploratory moves and systems in both realms, the epistemic and the artistic. Thus, Kubler's sequences, which appear as attempts at the solution of a form problem in, for instance, architecture or the fine arts, and which may span centuries, can directly be compared neither with experimental systems, nor with experimental cultures devoted to the solution of epistemic problems. Here, more often than not, old problems are not solved, but are replaced by newly emerging ones. But both Kubler's sequence perspective on the arts and the experimental system perspective on the sciences are undergirded by the common search for the "shapes of time" of their objects. It is also a search for new categories to make the different materializations of history visible and tangible. Kubler lent weight to this search when he wrote: "A net of another mesh is required, different from any now in use." And with respect to the traditional concept of style all too often invoked in this context, he added: "The notion of style has no more mesh than wrapping paper or storage boxes."[55] As a whole, the present book understands itself as a contribution to the elaboration of such a conceptual mesh.

In the end, the concrete and particular shapes of time in which human invention—that is, cultural novelty—realizes itself in history are to be understood as histories of things and are tightly tied to the respective materials and the options that emanate from these materials. Precisely because we are concerned here with an approach marked by the commitment to give the logic of the materials the place it deserves, the shapes of time that they carry with them and in which they unfold will distinguish themselves more or less significantly according to the artistic activities envisaged. The same holds, *ceteris paribus*, for the differentiations in the realm of the epistemic. It follows that what needs to be done is not to level out these differences, either within or across these realms, but instead to find a common ground on which they can be rendered comparable and conceptually recoverable in a productive manner.

Experimental Cultures

While chapter 6 focused on the shapes of time, chapter 7 deals with the specific shapes of the spatial expansion of experimental systems. As already mentioned in chapter 6, experimental systems can form ensembles, or bundles, that share one or more of the technical, epistemic-material, or social components of these systems. These ensembles can be described as experimental cultures.[1] The concept of *sharing* is important here. It essentially embodies a justification for speaking of cultures in this context. Certain aspects of experimentation that belong to this overarching level of analysis have been discussed in the literature using notions such as "styles of scientific thinking"[2] and "styles of scientific practice"[3] as well as "ways of knowing" and "ways of working."[4] The concept of culture emphasizes the material side of the interaction between experimental systems—an interaction whose importance becomes evident with the gradual development of these systems. Experimental systems and experimental cultures are thus reciprocally related. Contrary to the long-standing tradition that saw the concept of culture tied to the space of the symbolic,[5] our usage of the notion hones in on the materialities of the scientific working process. In addition, the focus on experimental cultures offers the perspective of envisaging longer time spans in the development of a particular science, or parts thereof, without having to depart from the level of scientific practice to embrace a perspective from the history of ideas.

The first part of this chapter looks closely at a particular experimental culture in the biological sciences: the test tube or in vitro experiment. Using a concrete example will allow us to unfold the concept of experimental culture historically and epistemologically. The example is chosen with caution, for experimenting in vitro underwent an unprecedented rise in the biochemically oriented practices of the life sciences in the twentieth century. In what follows, we will briefly discuss the most prominent of the diverse forms assumed by this kind of analysis of biological functions in the test tube over the course of the century. We will also consider the objections raised against this experimental view of life during the different stages of its development.

Early forms of experimentation in vitro were based on efforts to keep tissue cultures in artificial nutrient solutions and to measure their metabolic turnover on the one hand, and to develop cultures of single cells on the other. In parallel, the manipulation of cell homogenates started to expand. From the 1930s onward, the differential fractionation of cellular contents by means of a new instrument, the ultracentrifuge, moved into the foreground, and with that, the purification of particular cellular substances and organelles. After World War II the radioactive labeling of biomolecules added a completely new dimension to the tracing of metabolic processes such as the synthesis of nucleic acids and of proteins.

The second part of the chapter locates the concept of experimental culture in the broader discussion on the notion of culture in modernity. The ensuing sections outline the historiographical potential of working with the notion of experimental cultures. In contrast to a history of disciplines with its orientation toward institutions, a cultural history of the sciences such as that developed here underlines the epistemic dimension of the debate on scientific communities. Instead of following Thomas Kuhn in defining scientific communities by their shared paradigms,[6] we will try to characterize them in terms of shared experimental life forms in which researchers move and act.[7] A space for resistance and conceptual alternatives is structurally implicated in the niches and interstices of these communities. Consequently, these niches are understood not as anomalies, but as the driving moments of entire knowledge cultures.

Experimental Cultures: In Vitro

Biological experimentation in vitro had a deep and lasting influence on the development of the life sciences in the twentieth century. There are good reasons to argue that it was the test tube culture of biological experimentation that gave rise to the conjunctures between biology, physics, and chemistry that paved the way for the molecularization of the life sciences around the middle of the century. The linguistic distinction between in vitro and in vivo is itself a product of the development of physiology on its way toward biochemistry in the late nineteenth century. The *Oxford English Dictionary* quotes George Gould's *Illustrated Dictionary of Medicine, Biology and Allied Sciences* from 1894 as the earliest source for a definition of "in vitro." It states:

"In the glass; applied to phenomena that are observed in experiments carried out in the laboratory with microorganisms, digestive ferments, and other agents, but that may not necessarily occur within the living body."[8] This is ad hoc and unspecific, and at the same time, telling: it points to the origin of test tube experimentation in microbiology, and it plainly identifies the specter behind every manipulation of biological reactions "in the glass." Formulated as a question: is what we observe outside the body or the cell identical to what happens inside them? The question also points to a demarcation line. Around 1800, that line indicated the *ontic* division between the organic and the inorganic, thus becoming the starting point for biology as a science sui generis. Around 1900, the question revolved around the fixation of the *epistemic* conditions under which it could still be assumed that processes occurring within the organism could be made manifest outside the organism and thus subjected to analysis under these altered circumstances. At stake was the creation of test tube environments in which—always precariously—biological entities or processes could be subjected to measurement: entities or processes that were normally inaccessible to the scientific gaze and deeply hidden in the interior of the cell or the organism as a whole.

In what follows I will briefly outline the important stages and delineate the exemplary configurations in the historical development of biological test tube culture. A comprehensive historical account of this experimental culture has yet to be written. We shall see, however, that the experimental activity in vitro had a decisive impact on the reconfigurations of the life sciences over the course of the twentieth century. Not only the hybrid discipline of biochemistry was defined by this form of biological experiments;[9] the ensuing classical phase of molecular biology between 1940 and 1970 also rested on these presuppositions.

HOMOGENATES

The canonical history of modern biochemistry invariably begins with Eduard Buchner's report on the "alcoholic fermentation without yeast cells" published in 1897.[10] The preparation of extracts from both animal and plant materials for medical, technical, and nutritional purposes had been common practice from time immemorial—it dates back to prehistoric times. It was, however, not simply the manipulation of organic substances *extra corpore*, which was characteristic of the new test tube culture. According to Herbert Friedmann, what was really new was the claim and demonstration not only that tissue extracts contained so-called natural products (that is, organic *compounds* that were synthesized in tissues and organs and that could be purified and enriched outside the body), but that organic *processes* took place in the test tube. If we take seriously Friedmann's remark that "from now on, extract repeats or mirrors process,"[11] the transition from a living system to a test tube system is not simply a transition from biology to organic chemistry (or, to put it differently, from physiological processes to organic substances), but rather something like a replication of life—life under other conditions. This process of repeating or mirroring life is, however, permanently at risk of being distorted and thus confronts its proponents with the equally permanent question as to how far we still see *nature* in the mirror, and how we decide whether we still do that or not.

It was obvious that defining and stabilizing the fluid external milieu was the decisive task of a burgeoning in vitro culture—beginning with enzyme analysis—during the early twentieth century. A generation earlier, the French physiologist Claude Bernard had insisted on the essential importance of what he called the "fluid internal milieu," in which the components of the tissues were immersed and which he deemed indispensable for living processes. He was convinced that the "organic syntheses" characteristic of living beings could only take place in this milieu.[12] "If I see it correctly," he emphasized from hindsight, "I was the first to insist on the idea that there are really two environments for the animal, an *external environment* in which the organism is placed, and an *internal environment* in which the elements of the tissues live."[13] Bernard concluded that organic syntheses could only be investigated and assessed on—and in—the living animal. He became renowned and famous—not to say infamous—for the development of an in vivo physiology based on intricate, massively invasive surgical techniques and procedures in order to follow the transport and transformation of nutrients and other substances such as medicines and poisons in the different organs[14]—mainly in stray dogs roaming the streets of the Latin Quarter at that time. His experimental regime thus rested on sometimes extremely drastic interventions into the animal body, but they were intended to register the physiological transformations that went on within the living organism. Bernard therefore would have decidedly rejected a test tube experiment. Test tubes were only to be consulted retrospectively, to make chemical analyses of substances and to quantify them *after* they had been processed in the body itself.

Around the turn of the twentieth century, a new generation of physicians, biologists, and chemists began to search for an equivalent to Bernard's internal milieu that would allow them to turn the cellular interior inside out. This eventually gave rise to a widespread international community of researchers whose work revolved around the analysis of bioagents, in particular the structure and function of enzymes, those biological effectors par excellence.

Among them were August Fernbach and Louis Hubert at the Pasteur Institute in Paris, Søren Sørensen at the Carlsberg Laboratories in Copenhagen, and Leonor Michaelis and Maud Menten at the Charité in Berlin.[15] They all searched for conditions under which isolated enzymes—in particular those which were not secreted into body cavities by glands, but acted within the cells themselves—retained their metabolic function, or possibly even increased it.[16]

The creation of fluid environments not necessarily intended as a pure imitation of the cell sap—insofar as its constituents could be determined at all—but able to functionally replace the cytoplasm of the living cell, advanced to become one of the biggest technical challenges of test tube biology in the early twentieth century. At stake was the realization of a regime that was fundamentally different from organic chemistry, with its drastic syntheses at high pressures and temperatures under which biological processes were unthinkable. At stake was, to quote Friedmann again, "the duplication of a cellular chemical process outside an intact biological system."[17] In order to achieve this goal it was not necessary to reconstruct the complete "internal milieu" in the test tube. What counted was the creation of a functional environment in which a biological synthesis—or degradation, for that matter—could be represented in isolation in such a way that, ideally, its intermediate steps could be followed. A completely new culture in dealing with biological processes was emerging. Although it remained controversial for decades, it began to find its way into biological and medical laboratories around the world and to change their character completely.

Eventually, most processes that became tractable in the context of the new test tube biochemistry of the early twentieth century—with alcoholic fermentation playing a prominent role among them—emerged as complex metabolic cascades that stretched over several intermediate steps and involved the participation of not only one, but several enzymes or enzyme complexes and co-enzymes. Moreover, it became apparent that more often than not these enzymes were bound to subcellular structures such as membranes and organelles that had been poorly characterized until then and from which they could not be separated without losing their function. The question of whether the course and the activity pattern of a process in a test tube corresponded to the biological process within the cell could generally not be answered unequivocally. The experimental results had to be checked and revised repeatedly, but this permanent questioning kept the research process going. Many biologically oriented chemists—and chemically oriented biologists—thought it deeply problematic to replace traditional physiological chemistry based on whole cells and tissues with the new test tube biochemistry. Still, to mention

just one weighty example, in 1940 Richard Willstätter, who had won the Nobel Prize in 1915 for his work on chlorophyll, roundly condemned these kinds
of experiments. To quote one of his last publications on the enzyme systems
of sugar conversion, which he wrote with Margarete Rohdewald: "The post-
mortem changes occur in an uneven fashion, irregularly, without meaning.
The meat machine is the cliff of physiological chemistry. Observations on the
muscle homogenate or with extracts do not allow to draw straightforward
conclusions on anything about the living processes in the muscle."[18] Rohdewald and Willstätter passed the following judgment on Eduard Buchner's
experiments: "The option that macerated yeast sap would react with carbo-
hydrate in the same manner as living yeast was quite arbitrary and has to be
considered as refuted by now. The same holds for Buchner's press sap."[19]

Tissue and Cell Cultures

Given this unequivocal statement from a renowned physiology laboratory
four decades after Buchner's pioneering efforts, it comes as no surprise that
an alternative form of in vitro culture developed in parallel to test tube enzymology with its extracts. This was the extracorporeal cultivation of single
cells from multicellular organisms and, alternatively, of tissue parts. The latter
(that is, the technique of tissue slices) spread liberally into the laboratories of
physiology in the first half of the twentieth century. The result was that the
metabolic turnover of organs such as the liver could be assessed in a quantitative manner with the help of micro-measurements and in a controlled
environment.

In this context, the concept of culture assumes the specific meaning of
a cultivation. The concept was already used in this sense in bacteriology in
the late nineteenth century and was adopted in the early twentieth century
by those who studied isolated cells of higher organisms *extra corpore*. They
could follow in the footsteps of medical and biotechnological microbiology
with their methods of breeding bacterial "pure cultures" on the one hand,[20]
and protozoology with its efforts to cultivate unicellular eukaryotes such as
yeast, amoebae, or euglenozoae on the other.[21] The latter's efforts concentrated on the creation of an external milieu in which unicellular organisms
could be grown and induced to sexually proliferate.

A few examples in chronological order may suffice here to illustrate the
issue. Between 1906 and 1910, the embryologist Ross Harrison succeeded, first
at Johns Hopkins University in Baltimore, then at Yale University in New
Haven, to observe the outgrowth of axon fibers in explants of nerve tissue
from frog embryos in an in vitro culture. It turned out that the choice of

the substrate was the decisive factor for this success. The nerve cells began to grow when Harrison embedded them in coagulated lymph in a hanging drop.[22] Alexis Carrel, who originated the concept of "tissue culture," worked in a similar direction at the Rockefeller Institute in New York.[23] He was able to take human tissue and cells, in particular heart muscle cells, into culture. He even spoke, in that context, of the formation of "a new type of body in which to grow a cell,"[24] an extracorporeal vessel that nonetheless could assume the function of the body. In this context of medical research, isolated cells soon advanced to diagnostic and even therapeutic tools. Hannah Landecker has described this in detail under the suggestive title *Culturing Life*.[25]

In Germany, the cell physiologist Otto Warburg worked with intact cells in the test tube to elucidate the process of biological oxidation. In a collaboration with Otto Meyerhof from the University of Heidelberg, he had initially attempted to analyze the respiration processes in aerobic bacterial extracts. However, he soon realized that he only got results if he did not fractionate the ground cells further, and even these results proved irregular and sporadic. If he freed the extracts from the cell wall debris, the signal vanished altogether. Petra Werner remarked in her biography that Warburg consequently felt forced to "proceed directly and to use the whole cell as a 'test tube.'"[26] Warburg himself summarized his experimental investigations as follows: "Since experience teaches us that one cannot separate the catalysts of living substance— the ferments—from their accompanying inactive compounds, it suggests abandoning the methods of preparative chemistry and analyzing the ferments under their most natural action conditions, that is, in the living cell itself."[27] For this purpose he constructed a device named after him, the Warburg apparatus. It consisted of a temperature-controllable water bath, a manometer, a reaction vessel, recipient vessels and a shaking device. In the interwar period, it advanced to become the most important component of the laboratory equipment for an in vitro physiology based on tissue slices.

Warburg's contemporary Alfred Kühn, a pioneer in experimental physiological genetics, was among those researchers who remained skeptical toward the technique of destroying cells in order to analyze the reactions taking place within them in detail. Even in his later years, after the Second World War, he could not be persuaded out of his conviction, as we can see from a letter to his former coworker Ernst Caspari, who had been forced to leave Nazi Germany and had settled in the United States. After the end of the war Caspari reported to Kühn about his experiments, in which he had tried to investigate the formation of the eye pigments in homogenates of *Drosophila* flies instead of in the intact animal or at least in the intact organ. In his reply of December 1946, Kühn declared unequivocally: "I cannot believe that the whole synthesis that in vivo depends on numerous links of a gene action chain, and that not only polymerizes kynurenine, but also includes yet other components and lastly proceeds only if bound to protein granules, can occur in a tissue homogenate."[28] In accordance with Willstätter, Warburg, and many of his contemporaries, Kühn saw the intact organism as "a laboratory of a peculiar chemical regime,"[29] although he remained favorable to complementary biochemical analyses, as we have seen in chapter 4.

CELL FRACTIONATION

Throughout the early decades of the twentieth century, the only possibility of keeping whole cells or tissue slices as black boxes in the test tube and measuring their metabolic turnover by the resulting gases and simple chemical compounds with microchemical methods was the preparation of tissue extracts, either crude or centrifuged at low speed. With the advent of ultracentrifugation in the 1930s, physico-chemical institutes and medical and clinical research facilities endowed with the appropriate infrastructure began to subject such homogenates to high-speed centrifugation instead of working with the cytoplasm as a whole. The aim was to separate the cell sap according to the sedimentation velocity of its components into more or less well-defined fractions that could be analyzed either structurally, or functionally, or both. Differential centrifugation became the new buzzword.

Concomitantly, intensive efforts were focused on breaking up the cells as gently as possible so that the structures and complexes contained in the cytoplasm would be altered as little as possible. On the one hand, different physical methods and instruments were tried out, such as mixers, mills, mortars, vibrators, and ultrasound devices, as well as fine-grained aluminum or silicon oxide. On the other hand, tests were made with chemical procedures using solubilizers and detergents; even lytic enzymes were used. Their efficacy was assessed by means of the few standardized enzyme tests available in the 1930s for selected cellular functions. This example also demonstrates that the successful application of a new high-tech instrument is usually tied to and dependent on the optimization of rather low-tech or even trivial manipulations—a fact that is frequently neglected.

These experimental series belonged to the tradition of preparing cellular homogenates. The source of cellular tissue for most of these experiments was the liver of clinical laboratory animals such as guinea pigs, rats, and mice[30]— for two reasons. First, these animals were available in large numbers and stemmed mostly from inbreeding; they were thus, at least to some extent, genetically homogenous. Second, the liver is a highly active, regenerative tissue;

therefore, on the basis of its cytoplasm, robust activity patterns could be expected. In addition, it was necessary to explore the osmotic conditions under which the isolated cellular fractions—such as cell nuclei or mitochondria—retained their structure and remained functionally active. To do justice to this twofold requirement—gentle destruction of the cells and function-preserving fractionation—a broad spectrum of buffers, salts, and other soluble ingredients were tested. Given the lack of precepts to follow, trial and error predominated, which meant that these efforts went on for decades. Last but not least, it was necessary to fine-tune the physical conditions and parameters of the centrifugation procedure itself under which the cell sap could be separated into clearly distinguished fractions.[31] These conditions included the rotation velocity and the length of the run, as well as the form and the angle of inclination of the centrifugation tubes.

The Belgian physician and cytologist Albert Claude, who after he had obtained his medical degree went on to work at the Rockefeller Institute in New York, counts as one of the pioneers of cell fractionation. In the late 1930s and early 1940s he had introduced high-speed centrifugation in order to isolate the active principle of Rous sarcoma from chicken tissue. Soon, the tumor agent revealed itself as embedded in a co-sedimenting cellular component of unknown composition and function, and the attempt to separate them from each other seemed hopeless. However, the homologous sedimenting cellular component from non-infected cells was accessible to quantitative analysis and measurement, and Claude turned to its determination instead. In doing so he laid the foundation for in vitro cytology.[32] Starting from a complete homogenate, and faithful to the principle of not disposing of any of the resulting fractions and their components, Claude was able to define four fractions that were sufficiently separated from each other: a fraction sedimenting at low speed, containing cell-wall fragments and nuclei; a fraction containing the mitochondria, including the respiratory ferments; a microsomal fraction whose particles were too small to be seen under the light microscope; and a non-sedimentable supernatant containing the proteins and nucleic acids remaining in solution as well as the low-molecular residue of the cytoplasm.

Claude was not completely alone in developing this kind of in vitro cellular analysis. He was able to draw on the work of Robert Bensley and Normand Hoerr at the University of Chicago, who were working on the in vitro representation of mitochondria. In Europe, the group around Jean Brachet from the Free University of Brussels competed in the isolation, fractionation, and biochemical characterization of microsomes from yeast cell extracts.[33] But although Rollin Hotchkiss, George Hogeboom, and Walter Schneider at the Rockefeller Institute had decidedly refined their test battery on the basis of an assessment of enzyme activities—calling their procedure "biochemical mapping"[34]—neither group succeeded in ascribing an unequivocal biological function to the microsomal particles. It would be another decade before this was finally possible.

RADIOACTIVE TRACING

A more specifically targeted, functional in vitro analysis of the microsomal fraction only entered the realm of the possible with the introduction of another technology: the radioactive labeling of biomolecules, which played a role in metabolic as well as genetic turnover. This advance signaled a new direction for practically all of biological in vitro research. I already adduced this technology in chapter 1 above to illuminate the process of the experimental generation of traces. After the development of counters that were sensitive enough to register the decay of radioactive hydrogen (^{3}H) and radioactive carbon (^{14}C) as well as radioactive sulfur (^{35}S)—isotopes that became available in larger amounts as byproducts of atomic reactor technology[35]—the method of radioactive labeling advanced to something like a biochemical electron microscope. The introduction of radioactive substances into the laboratory cultures of biology not only had an impact on the experimental systems in which they were used (that is, the experimental culture of in vitro cytology that was on its way to becoming a basis for molecular biology); it also affected laboratory architecture and laboratory life as a whole. The capacity to measure radioactive traces of minimal intensity in biological probes depended on the possibility of avoiding a contamination of the laboratory environment in which the experiment took place. Consequently, it compelled the establishment of a new laboratory regime with isotope chambers and double door systems. From an epistemic perspective, however, it also had direct consequences for the design of the experiments on a new scale, in terms of both sensitivity and size.

These few remarks may suffice to demonstrate that the technology of radioactive tracing cannot be reduced to an instrument—such as the liquid scintillation counter—or to a substance—such as an isotope. It was a technical ensemble, a ramified research technology that with its different facets impregnated and penetrated the whole culture of in vitro experimentation. The decisive aspects of the tracer culture can be summarized as follows.[36] First, the tracer technology realized an indicator principle that offered the possibility of transcending the quantitative barriers of chemical micro-analysis. The massive damping that biological functions generally suffered in the test tube was compensated by the sensitivity of the technique. In other words, what the cell sap lost in terms of biological activity by its preparation, it regained by the addition of radioactivity. Second, the substances no longer had to be isolated and purified in order for researchers to be able to quantify them. Instead, measurements could be carried out with a complex background. Third, the labeling technique itself became a driving force for the development of new radioactive counting technologies such as liquid scintillation counting. Its integration into the experimental systems of biological chemistry not only massively widened their scope, but also opened up new options for setting up experiments, such as extended kinetic measurement series. Fourth, and finally, the technology constituted something like a material mediator of the knowhow from different fields of expertise: biology, chemistry, physics, and engineering. It was a technology whose very design—the confluence of isotope physics, the chemistry of liquid scintillation and labeling of biomolecules, electronics, and biological sample preparation—required the conjunction of expert knowledge from these different areas. It also transcended the traditional disciplinary boundaries, representing a nodal point that plainly demonstrates why these boundaries tended to lose their importance accordingly.

Taken together, the combination of differential cell fractionation and radioactive tracing led the culture of in vitro biochemistry, with its structurally and functionally oriented experimental systems, into the domain of the cellular synthesis of macromolecules, and thus into the center of molecular biology, which underwent an unprecedentedly explosive development after the Second World War.[37] For a long period this biochemical and experimental foundation of molecular biology remained overshadowed by the luster of a handful of illustrious physicists who had invaded the field in the postwar period. It only slowly received the historiographic attention it deserves.[38]

The creation of an extracellular environment for the analysis of the intracellular reduplication and expression of genetic information is a prime example of the kind of test tube biology that became the norm in the first three decades after the Second World War. Toward the end of this period, in the 1970s, researchers succeeded in breaking the high-molecular complex of the bacterial ribosome in the test tube down into its components, and reassembling it again under appropriate buffer and incubation conditions as a fully functional particle. The in vitro reconstitution of functionally active ribosomes, along with the analysis of the partial reactions of the complex cycle of peptide bond formation in protein biosynthesis, can be seen as the apotheosis of test tube molecular biology.[39] Since then, gene technology, itself a result of the development of molecular biology, has been on the road into the cell and has taken possession of it again. With its molecular tools it has left the test tube and has returned into the intact cell as the space of its experimental interventions.

From the perspective of experimental cultures as adopted here, two preliminary historiographical consequences can be drawn that will occupy us in more detail in chapter 8. To begin with, experimental cultures are themselves transitory ensembles. Their epistemically productive lifespan is limited. They can become dominant, their different manifestations can coexist for some time, but they can also be marginalized and disappear altogether. To stay with our example, fractionation of the cell sap replaced mere homogenization, but cell cultivation coexisted, first with homogenization, and then with fractionation. Today, cell cultures proliferate again,[40] while the other forms of in vitro experimentation are on the decline. What is more, experimental cultures also allow for the establishment of connections between different fields and thus for the subversion, if not surmounting, of the institutional academic barriers of disciplines. Even if they cannot do away with these socially petrified institutions, they regauge and reconfigure their limitations. From the perspective of the entire twentieth century, the cultures of in vitro experimentation had an enormous impact on the multilayered traffic between biology and medicine, to take just one example. Today's biomedicine is largely a result of these reconfigurations.[41]

Culture as an Epistemological Concept

The concept of culture occupies an extended horizon of meanings.[42] In its broadest sense, its modern use is bound to the categorical distinction between things made by humans and naturally occurring things. The Hungarian philosopher and founder of cultural sociology, Karl Mannheim, convincingly showed that this distinction itself had to emerge and solidify historically, and in its fully-fledged form—as a sharp separation—became the core of the cultural self-understanding of modernity. For us modern people, according to Mannheim, "being and meaning, reality and value have been broken up and disintegrated for our experience," and it is only through such a disintegration in everyday life that "the designation of culture as non-nature became genuinely concrete and internally consistent." Henceforth, culture meant that which could be understood as engaged in a "mental-historical trajectory." It was only its acceleration that made a perception of the distinction possible. In distinction, "the nature which makes up the contrast to the modern 'culture,' its correlate, then, is something . . . itself devoid of meaning or value. It gathers together the accumulation of all the qualities which do not pertain to the cultural. Nature is thus that which cannot be penetrated by the spiritual, that which is indifferent to value, that which is not subject to the spiritual-historical course of development."[43] It is thus a double negation that underlies this circular determination—culture as non-nature, nature as non-culture.

In his book *The Natural Contract*, Michel Serres sharpened this diagnosis in a forceful manner: "Take away the world around the battles, keep only conflicts or debates, thick with humanity and purified of things, and you obtain stage theater, most of our narratives and philosophies, history, and all of social science: the interesting spectacle they call cultural. Does anyone ever say where the master and slave fight it out? Our culture abhors the world."[44]

Against this horror, Serres's call for a natural contract can only be understood as a paradox under the regime as described. And yet, nothing less is required in a situation in which "nature manages to penetrate into the pegged and walled realm of culture. The stone falls from the firmament into the city, from physics into law."[45] In this situation, Serres directs his call of duty to the sciences in particular, for they have aided and abetted that irruption of the real.

It is a comparable paradox to talk about cultures of experimentation or cultures of the natural sciences whose aim is supposedly to understand a nature completely free from meaning and culture. But the paradox also points ahead to a view of the sciences that pretends to locate itself beyond the modern dichotomy. It transports and reflects the increasingly urgent effort to locate the sciences—the sciences of nature—not simply and without further caveat on the side of their objects, and with that, to reclaim them as culturally invariant. In the discourse of the natural sciences on their own procedures and in the public image of themselves that they have long disseminated, this happened—and still happens—again and again. Against this tendency, the arguments for *Science as Practice and Culture*,[46] for *Epistemic Cultures*,[47] and for scientific knowledge cultures[48] aim to perceive and treat scientific knowledge as a cultural phenomenon in all its historical changeability.

That goes a decisive step beyond Mannheim. For him, and in his critical reflection on the modern dichotomy of nature and culture, it was still beyond question that knowledge about "the historical-cultural reality" alone was to be seen as a product of its respective historical-cultural embeddedness. In contrast, he saw natural scientific knowledge as "tied to its own history, only inasmuch as later knowledge may be said to presuppose all prior scientific results as necessary premises."[49] Mannheim thus attributed a kind of historicity to natural scientific knowledge, but a purely internal history free of culture.

Pierre Bourdieu once characterized this dilemma—the claim to objectivity versus cultural boundedness—as the inescapable "double face" of scientific knowledge. In *Méditations pascaliennes* he described the dilemma as follows:

> While it forbids one to move fictitiously beyond the uncrossable limits of history, a realist vision of history leads one to examine how, and in what historical conditions, history can be made to yield some truths irreducible to history. We have to acknowledge that reason did not fall from heaven as a mysterious and forever inexplicable gift and that it is therefore historical through and through, but we are not forced to conclude, as is often supposed, that it is reducible to history. It is in history, and in history alone, that we must seek the principle of the relative independence of reason from the history of which it is the product; or, more precisely, in the strictly historical, but entirely specific logic through which the exceptional universes in which the singular history of reason is fulfilled were established.[50]

Once Again: Experimental Cultures

Whether one holds with the cautious Bourdieu or prefers a more radical variant of historical epistemology, we are obliged to enter into a deeper discussion about the concept of culture for the characterization of the sciences and their development. At the same time, this poses a challenge to the modern separation of nature from culture. As a starting point I actually prefer a descriptive position. I have proceeded on the assumption that, taken together, ensembles of interrelated experimental systems can be seen as experimental cultures. If experimental systems can be characterized as the smallest functional units of modern experimental research, localized in particular laboratories, experimental cultures present ensembles of such systems bound together and able to communicate with each other in one way or another—beyond the mere reception of scientific results recorded in writing. Here again, the concept of sharing comes into play. If I see it correctly, such ensembles should fulfill at least three conditions of sharing; whether they are sufficient to establish such a culture may remain an open question for the time being. First, we require a certain overlap of the techniques on which such clusters of experimental systems are based and which come to be used in them, mostly in various combinations. Experimental cultures thus share research technologies. Second, there must be an exchange of matter between the systems that form an experimental culture. In other words, their epistemic objects must exhibit overlaps. Third, experimental cultures are distinguished by the circulation of research personnel. The know-how and skills that the members of one of these systems have acquired can thus be transferred to others. Because of this flow there is a permanent input that creates a link between the two options of compatibility and a view from outside. The transition from one experimental system into another can be successful because of compatibility. But it also evokes certain effects of estrangement. That leads to moments of heightened attention, inducing constellations that encourage the occurrence of epistemic events.

Beyond formal and public scientific communication, networks of experimental cultures are therefore distinguished by a triple circulation: of technologies, of objects of investigation, and of skills tied to mobile persons. With that circulation in place, they exhibit an epistemic cohesion that is decidedly different from what we associate with the concept of a scientific discipline that is usually defined by forms of institutionalization and ways of passing on a specific corpus of knowledge. In contrast, the concept of experimental culture focuses on the research process, that is, the process of gaining knowledge in its informal dimensions. Disciplines are usually—although not always—defined ontologically by their range of objects. Experimental cultures, though, define themselves epistemologically by their attitude and characteristic approach toward a range of objects. The in vitro culture of biological experimentation of the twentieth century is exemplary in this respect.

Gaston Bachelard argued in a similar direction in requiring that an epistemology concerned with the actual development of the sciences take seriously the dynamics of regionalization of the latter. According to Bachelard, the modern sciences create spaces that are distinguished by "kernels of apodicticity."[51] Despite their pretension to unconditionality and their occasionally esoteric exclusivity, they appear both temporally and spatially confined. Each of these kernels requires its own epistemological attention. In his *Philosophy of No*, Bachelard pleaded in this respect for a "philosophy of epistemological detail": "A philosophy of epistemological detail needs to be founded, a *differential*, scientific philosophy which would constitute a counterpart to the *integral* philosophy of philosophers." In other words, a historical epistemology. "This differential philosophy would be responsible for following the gestation of a thought."[52] Or, to use the words from *Applied Rationalism* published ten years later: "We have to attain a concrete realism that is solidary with the particularity and precision of the experiments on which it draws. Such a realism has also to be sufficiently *open* to receive new determinations from exactly these experiments."[53] An epistemology that claims to do justice to the dynamics of scientific work at the boundary of the unknown must be as mobile and ready to risk itself as the sciences it tries to understand. "On the whole, it is a question of realizing in depth, philosophically, the experience of novelty. One cannot attain this profound renewal without a disponibility of the philosophical spirit."[54]

In this context, Bachelard also spoke of "cantons," regions or quarters in the "city of knowledge."[55] For him, such cantons or districts represented islands of a scientific culture with their own codes, semantics, and forms of emergence. He himself used the concept of "scientific culture"[56] for them, though not without attributing a peculiar meaning to it. He defined a scientific culture as the "access to an emergence"—*une accession à une émergence*.[57] Scientific cultures in Bachelard's sense are thus specific epistemic milieus in which the generation of new knowledge can happen, in which unprecedented epistemic events can occur. They are cultures of innovation. If, as Bachelard sees it, the artistic act of the creation of new forms is more firmly stamped by the individual, the scientific act of the creation of new knowledge is decidedly tainted by the collective. Scientific "emergences" are, says Bachelard, "effectively constituted socially."[58] That also means, however, that they take the shape of cultures in the sense of a "union of workers of proof"[59]—a community that deals with the phenomena of its epistemic interest in a rather specific form, but one shared by all its members. These cultures keep the process of supplementation of phenomena going—the process of developing novel approaches and the accompanying concepts that are characteristic for each of the particular research cantons. The more closely these districts are circumscribed, the easier it is to modify or alter conventions, forms of measurement, ways of description, and grids of classification, or to export them into other research areas. Regionalization creates epistemic flexibility. Bachelard refused to recognize a deplorable loss of synthetic gaze in this fragmentation of the cultures of contemporary research. In fact, he saw it as a precondition for the unprecedented fertility of the sciences of his time.

Besides the three circulation criteria described, there are altogether five aspects in the characterization of the sciences that form the basis for the concept of cultures of knowledge. The first concerns the working process of the sciences and the conditions under which it is articulated. It merits proper attention because the creation of new knowledge depends on its creative reconfigurations. Even in its advanced forms this working process still goes along with artisanal know-how. Second, the concept of culture points to the fact that the sciences generally are social, collectively constituted working formations. On the one hand, this point includes the laboratory as a collective space of knowledge generation; on the other, it refers to the peculiar handling of materials, their cultivation, their thingness that has to be seen in its own right as a collective knowledge reservoir. Third, we must stress the use of the plural "cultures." The concept has to be understood here as a combat term, for this pluralization strategically subverts the talk about the two cultures. The sciences are no longer to be divided into a genuine, real, hard, leading culture, and an inauthentic, metaphorical, soft, and less important culture. It has to be realized that, in the realm of the sciences of nature, as elsewhere, there is an irreducible plurality of objects and matching procedures that break up *the one* science from within and at the same time lead to the formation of intercultural zones in which new hybrid forms of understanding are tested that can in turn lead to further productive differentiations or amalgamations.[60] Following a suggestion by Susan Leigh Star and James Griesemer, the objects that are at stake and negotiated here can be called "boundary objects."[61] Fourth, it is a hallmark of experimental cultures that they are not only passively subjected to historical change, but that, as accesses to an emergence, they also provoke that change and make it possible. In the sense of an immanent transcendence, the historical is at work in the deepest core of scientific cultures. Fifth, and finally, the concept of culture in its application to the sciences is also meant to open the perspective of bringing the sciences as specific knowledge cultures into contact and in relation to other cultures of knowledge, such as cultures of the arts and technical cultures, among others. For the arts, this is attempted in more detail and from a diachronic perspective in chapter 6.

In conclusion, cultures of experimentation can be described as historically bounded forms of treatment of their respective research objects that deploy their power within limited time periods. In many cases it is only by the creation of such specific cultures that a particular phenomenon of one of these objects can be revealed and become accessible for research. Experimental cultures draw life from the phenomeno-technical events that Bachelard called "emergences," and at the same time they bring them into being. They are working environments in which new knowledge can take shape. Like experimental systems, they are—as ensembles of such systems—structures that must be explored and represented in all their concreteness and historical contingency. They are structures with a phenomenology in which epistemic, technical, and social moments are inextricably intertwined. In that sense, they draw life from concretizing, not from abstraction. Particular experimental cultures can dominate whole epochs in the development of a science. The in vitro cultures of biological experimentation whose contours I have sketched here played exactly this role for the emergence of molecular biology around the middle of the twentieth century.

8

Knowing and Narrating

When the modern sciences reflect on their own activity—that is, when they reflect on what distinguishes them from other discourses and makes them special—they let the things themselves appear written in intelligible letters; we only have to learn to decipher them, so the story goes. In this chapter on knowing and narrating,[1] I would like to give a particular twist to this traditional image by following it up and subverting it at the same time.

According to the metaphor of the legibility of the world, the letters in which the book of nature is written are inherent in that nature. We need only find a way for them to be revealed. As Hans Blumenberg has shown, this image accompanied not only the creation stories, but also the sciences from the early modern period up to the present.[2] From Galileo Galilei's mathematical vision of the book of nature, to the graphical method of nineteenth-century physiology with its automatic writing devices, up to the molecular genetic four-letter universe of our time, it was repeatedly recalled in a wide diversity of forms.[3] If we follow this image, the demarcation criterion of a discourse that lays claim to scientific authenticity is to allow the things to articulate themselves according to their own grammar and lexicon. If this succeeds, the scientific discourse would appear transparent and unadulterated by the media through which it is represented. Against this background, the important question would not be whether scientific texts narrate or not. Their scientific validity would not be due primarily to the fact that they proceeded descriptively or according to the mode of a hypothetico-deductive explanation, or that they followed similar, more or less equivalent epistemological distinctions, such as that between explanation and understanding, between the nomothetic and the idiographic, between knowledge of nature and knowledge of history. Instead, the fact that they have *another author* is what would distinguish them from the many histories, whether invented or true, that we tell ourselves about anything and everything in the mode of history or fiction.

The Question of Authorship

To pose the question of narration and narrativity from the perspective of scientific knowledge thus means not simply to search for its objects and their concatenations into stories, but ultimately to search for the *subject* of the sciences. Sooner or later, trying to answer this question, we are faced with an alternative that can be formulated as follows: Either one trails the line just sketched and lets the scientists disappear behind the transcendence of the divine or secular order of things, in which case the things announcing themselves in their own idiom become the subjects themselves. Or one opposes this kind of hypostasizing objectivism and filters the scientific discourse about the order of things into the stream of all those stories that *we humans* tell ourselves, in which case the sciences present themselves according to temperament as *one* big or many small stories. In either case, they lose their epistemologically singular character. This particular demarcation line appears impossible to avoid. It has left its mark on the debates in philosophy and history of science, and it continues to haunt them to this day.[4]

Research Machines

In *Toward a History of Epistemic Things*, I tried to show that in the modern sciences the experimental order of things realizes itself in a dynamic process condensing in experimental systems. They can also be addressed as research machines.[5] In chapter 6 above I expanded on their temporal aspects. Now I hope to be able to convince the reader that a closer look at such systems and their dynamics opens up a perspective that points beyond the dichotomy between self-disclosure and constructivist relativisms regarding scientific knowledge. I hope that such a praxeological approach, which involves the definite transition from a view of science as an epistemically privileged—or even not so privileged—*language* to a view of science as an epistemic *practice*, offers a possibility of escaping this dichotomy, or at least making it porous. We can see experimental systems as techno-epistemic environments that allow us to transform things of scientific interest into phenomena that can be investigated materially. In any science that has developed to a certain extent, such contexts are indispensable for the manifestation and analysis of epistemically interesting phenomena.

For the experimenters, this context presents itself primarily as an instrumental one: as a *conditio sine qua non*, a necessary platform for their work. On
second thought, however, it quickly becomes evident that the instruments that
go into this context consist of what can be addressed as "sedimented" knowledge (to echo Edmund Husserl[6]) or "reified" knowledge (in the words of Gaston Bachelard).[7] In their respective analyses, Husserl placed more emphasis
on the conceptual, while Bachelard stressed the instrumental side of the matter. Research technologies—procedures as well as apparatus—are thus components of knowledge environments in a materialized form. We can regard
them as collectively authorized vehicles, as knowledge transmitters that, in
their orientation toward an epistemic object, represent something like trans-
subjective generators of events. They are machines for the generation of epistemic differences. Borrowing the term "geno-differences" coined by Wilhelm
Johannsen to characterize gene actions, we could describe them as "epistemo-
differences."[8] Sedimented knowledge present in such research configurations
in a wired form is thus equipped with the power of differential action.

What does that mean more precisely? The case study on which my above-
mentioned book is based is a good occasion for a more detailed discussion on
the kind of action power that is at stake here. Immediately after the Second
World War, a small group of young researchers gathered around the oncolo-
gist Paul Zamecnik at Harvard Medical School and Massachusetts General
Hospital in Boston to tackle the problem of cellular protein biosynthesis
experimentally. The promise of the emerging new technique of radioactive
tracing had boosted their hope and confidence in their ability to follow the
biosynthesis of proteins in vitro and to establish a system on the basis of cellular tissues and, eventually, homogenates. The starting point was the hope
to distinguish between malign and healthy tissue on the basis of their different protein synthetic activity—as an entry point for a better understanding
of the physiology of cancer. At the time, the predominant theoretical view
was that protein synthesis could be understood as a reversal of proteolysis
(the degradation of proteins by digestive enzymes). The latter had already
been characterized by classical enzymology. But things would turn out otherwise. The difference in the protein synthesis activity between normal and
malign tissue emerged as insignificant. The result was that, over the course
of a decade and a whole series of experimental displacements, the idea of an
enzymatic peptide synthesis was replaced by the concept of a bioenergetic
charging of amino acids, followed by their condensation into chains alongside a matrix. The group was diverted from standard cancer research into
bioenergetic chemistry. When the integration of an ultracentrifuge into their

working system revealed a hitherto unknown new molecule that turned out to be a hybrid between nucleic acids and an amino acid, a mediator of the matrix-driven protein biosynthesis was identified. As researchers followed the trail of this molecule, the experimental protein synthesis system became the focus of emerging molecular biology in general and molecular genetics in particular. Finally, it became the arena for the deciphering of the genetic code. In the course of fifteen years, three decisive turns had happened, none of which the actors had been able to foresee. Each of these turns appeared as a contingent surprise, and each of them led to corresponding reorientations and reconfigurations of the experimental system. And yet, the whole trajectory was originally, and remained, based on an underlying material continuity without which the turns themselves would not have been possible.

Narrativity

Experimental systems embody and realize a narrative structure. It is based on the machinery that fills the space between the person wanting to know and the thing to be known, and it determines the dynamics of its generation of events. If the generation of knowledge is a subject in its own right at all, it is because of the historicity of the process that is being driven by the dynamics of these middle grounds. This machinery appears as a collective carrier beyond any immediate and unmediated confrontation of an *ego* with the world. Bachelard once characterized the situation like this: "It is no longer about confronting a lonely spirit with an indifferent universe. From now on, one has to act from within that center in which the cognizant spirit is determined by the precise object of its effort to know, and from where the former in turn directs experimentation with ever greater precision."[9] That implies that a phenomenologically oriented epistemology striving to capture such a movement has to understand itself as a permanent historical task. The late Edmund Husserl formulated this insight as follows in his remarks in the "Origin of Geometry," a key text of historical epistemology: "Certainly the historical backward reference has not occurred to anyone; certainly theory of knowledge has never been seen as a peculiarly historical task. But this is precisely what we object to the past." And he added: "The ruling dogma of the separation in principle between epistemological elucidation and historical, even humanistic-psychological explanation, between epistemological and genetic origin, is fundamentally mistaken."[10] If one takes this coupling of history and epistemology seriously, it becomes impossible in principle to separate the things of knowledge from the conditions of their emergence.

In a genuinely historical epistemology therefore, the "vital movement of the coexistence and the interweaving of original formations and sedimentation of meaning" can only be grasped as a whole, no longer neatly separated from each other.[11] László Tengelyi speaks in this context of a "conversion of Husserl's phenomenology in its latest phase of development," whose questions can no longer be encompassed in a purely intentional fashion.[12] Or concisely, to quote Jacques Derrida, "intentionality" becomes "traditionality."[13]

In the historical movement of knowledge generation, in which the sedimentation of meaning increasingly takes the form of research technologies, the experimental order of things undergoes a permanent realignment. The result is that experimental systems not only narrate stories but also change them. In doing so, they not only open up toward the future, they also make their past appear in an ever-changing light. With every substantial new acquisition, the meaning and sense of the sedimented knowledge is subjected to a reconfiguration. The characteristic openness of scientific knowledge toward its future modification also provokes a redefinition of what lies in the past—a movement that is inscribed in all historicity, but comes fully into its own in the space of the scientific. Thus, not only scientific questioning ahead, but also the "return" or "regressive inquiry" (die Rückfrage[14]), that genuine gesture of all things historic, is grounded in repetition and difference.[15]

This retroactive force of experimental questioning into the future that also takes possession of the past was again at work in the experimental trajectory just described. From the late 1950s to the late 1980s, Paul Zamecnik was repeatedly invited to summarize the results of his laboratory. Looking through these reports one after another, it is astounding to see how thoroughly the presentation of the turning points of this research trajectory changed over a quarter century. In the late 1950s, when the work of the group was just beginning to enter the limelight of the nascent field of molecular biology, Zamecnik presented a story that was largely operational in its terms and reflected the techniques that the group used to dissect the investigated biosynthetic process into its constituent parts.[16] Amino acids were "incorporated" into proteins, reflecting the evidence that in these experiments, the newly added radioactive amino acids became closely associated with protein. "Microsomes," and later on "ribonucleoprotein particles," were identified as the sites of protein biosynthesis. The former term related to a cytoplasmic fraction that could only be sedimented with high speed in the ultracentrifuge, the latter to the chemical composition of a purified form of these particles. The ribonucleic acid that could not be pelleted at such high rotating speeds of the centrifuge and that, to the surprise of the investigators, became charged with amino acids, was called "soluble RNA." And so on. The terminology remained attached to the reified part of the experimental system and superimposed the performative idiom of classical enzymology with its "intermediate products"—and so on. Conceptually the group tried not to commit itself prematurely. Twenty years later, the report sounded completely different: from the late 1940s on, the laboratory had been concerned with the exploration of the "ribosomes." It discovered "transfer RNA" around the middle of the 1950s, and with that, added a decisive building block for the understanding of the "language of the genes and proteins"—that is, the process of the "translation" of the genetic message into its biological function, as it took shape in the late 1950s.[17] Now, the history of the laboratory was a molecular biological story from the outset. In the light of what happened in later years, the origin appeared as the beginning of something that could not even have been conceived of back then. The course of the experimental process had also caught up with the meaning of its own history. To quote Bachelard: "Everything has to be taken up again, if it becomes necessary to envision completely new relations."[18] *Rational memory*" had taken possession of "*empirical memory*."[19]

Contingency

Channeled contingency thus turns out to be the essence of the historical. This means not only that it demands that "regressive inquiry" be constantly repeated, but also that it conditions the events lying ahead. Science, as research, is an epitome of historicity. To quote Bachelard once more: "The history of the sciences appears, then, as the most irreversible of all histories."[20] History lives from events; the history of the sciences lives from and with "*epistemological acts*."[21] Without given boundary conditions, however, an incidence cannot present itself as an event and be perceived as such. In this situation difference is always already embedded in repetition. Without events, without the creation of singularities, there is no historical process, and without repetition there are, in turn, no events. The historiography of the sciences has to occupy itself with this particular kind of epistemological prolapse.

Following Hayden White, who refers to a discursive distinction drawn by Emile Benveniste, the historian has the choice between "a historical discourse that narrates and a discourse that narrativizes." The former "openly adopts a perspective that looks out on the world and reports it," that is, understands itself explicitly as an account given by a historical narrator. The latter is a discourse that "feigns to make the world speak itself and speak itself as a story."[22] Benveniste characterized this latter position even more sharply: "No one speaks here. The events seem to narrate themselves."[23] "Narrativization" in this sense, and "historicality" in the sense of a mode of the temporal, of a structured but event-driven process, as Paul Ricoeur understands it, are thus inextricably intertwined.[24]

. . . IN LITERATURE

This narrative phenomenon in its double structure, a narrational and a historical, is at stake here. Our task is not only to determine a linguistic structure, but to lay bare the roots of this double structure on the example of the historical shape of scientific research. This brings processes into focus which appear to offer the basic conditions of possibility of such forms of narrativity. Nevertheless, a comparison with the production of fictional texts can serve as a starting point. In the space of literature, like that of history, no story that deserves the name can do without offering surprising turns in the framework of a narrative exposition. And here as well, its past appears in a new light at every turn. A really good story gives rise to these unexpected twists through the exposition itself. The structural homology between factual and fictive narrativization on which Hayden White insisted and with which he struggled himself, not without pain, has at its core the relation between confinement and transgression, between repetition and difference.

In his book *Grains et issues*, Tristan Tzara, one of the founding figures of Dadaism and surrealism, described his view of a good literary story as follows—in a passage with the spoken title "Experimental Dream":

> Thus, the story follows, and spreads across, the frame of a logical development that reduces itself to an account of successive facts, but leaving an irrational and lyrical remnant open for discovery. This, in turn, overflows the vessel intended for it, and at times engulfs and floods the base, the foundation, the traditional scaffolding of the story. It is a lyrical superstructure whose elements are derived from the base structure and which, once it is realized, impacts back onto that structure from the heights of its new power. Occasionally, its force intensifies to such an extent that it undermines the meaning of the structure, corrupts it, abolishes it, annihilates it in its essence.[25]

In this way, and in the realm of literary production, Tzara sketches a scaffold that allows for "bringing forth new events not foreseen by the original plan."[26] It is evident that the historical process is the inspiration behind the description of the literary process. The writings of Tzara—poet, writer, political activist, mastermind of surrealism in France in the 1920s and 1930s, and co-editor of *Inquisitions*, the organ of a "group for the study of human phenomenology"—served Bachelard not least as a foil for what he perceived as the scientific "surrationalism" of his time.[27] Conversely, he understood the experimentalism of the contemporary sciences as a resource for poetological reflection.[28]

. . . AND IN THE SCIENCES

In the realm of the sciences, or more precisely, of scientific research, experimental systems can be seen as the structures able to generate such contingencies. They create a space for events that challenge existing knowledge, ensuring epistemic surprises that are more than the dubious intuitions of a moment of luck, and of a different nature. Experimental systems are arrangements that bring forth history through their temporal order, and at the same time write stories through the permanent displacement in meaning that they cause. The history and the stories that the sciences thus generate are the result of these peculiar difference machines that owe their own genesis to the development of the modern sciences. Consequently, the historiography of science has to pay particular attention to them. They form the kernel around which a poetology of research can be built with a center that is not rhetorically but rather praxeologically constituted.

Poetology of Research

Michael Polanyi—who began his career as a physical chemist in Berlin and continued it later as a philosopher of science in Manchester—left us a stance that can serve as a starting point for such a praxeologically oriented poetology. Marjorie Grene quotes one of its versions from a lecture manuscript in her book *The Knower and the Known*.[29] It is a statement on what we can call the *research situation*. A published version reads as follows:

> We make sense of experience by relying on clues of which we are often aware only as pointers to their hidden meaning; this meaning is an aspect of a reality which as such can yet reveal itself in an indeterminate range of future discoveries. This is, in fact, my definition of external reality: reality is something that attracts our attention by clues which harass and beguile our minds into getting ever closer to it, and which, since it owes this attractive power to its independent existence, can always manifest itself in still unexpected ways. If we have grasped a true and deep-seated aspect of reality, then its future manifestations will be unexpected confirmations of our present knowledge of it.[30]

Such a perspective on the objects of research exposes a series of facets that, taken together, play around the poetological character of research in a succinct way. What is immediately striking is that the quote begins with the material conditions of the research process. To be sure, research implies the mind of the researcher; it has to dispose of a certain kind of sensibility to which I shall return shortly. But its results do not spring out of this mind like a sudden illumination. Rather, the mind of the researcher appears to be bound into a particular constellation. In the context of an experimental procedure that allows things to be transformed into objects of research, acquaintance is made with matter. Without that mediation, there is no research process—it relies on the necessary experimental setup. The scientific mind—*l'esprit scientifique*, to use Bachelard's words[31]—presupposes this exteriority which not only chafes at it, but can allow it to take shape. Without this permanent challenge—this exteriorization—it would not be capable of doing much more than revolving around itself and perhaps, at best, getting something out of its own fluctuations from time to time.

The second observation is that Polanyi endows the "things" around which research revolves with a kind of agency—that faculty to "manifest [themselves] in still unexpected ways." They possess the potential to surprise the researchers: to frustrate their expectations, to dwarf their capacity for foresight by their own richness of disclosure. It is difficult to know who is the master in this game and who the slave. Experimenting means organizing the process of knowledge generation in such a way that the things are always allowed the penultimate word, a word that generally cannot be anticipated and that is never definitely the last word. Clever experimentation leaves the door open for unprecedented things.

The third point is that the agency of things implies resistance, resilience against arbitrary access. Their experimental malleability is restricted; we cannot simply do what we want with them. Therefore, we cannot talk about research work unconditionally as a pure activity of construction, in the sense of either a technical or a social constructivism. Experimenting, however, does not amount to discovering in the usual sense of the word either (that is, merely exposing to view). Experimentation is about *manifestation*. Manifestation is a wonderful word for denoting a movement that lies in between. The word implies that the movement it describes needs the experimenter's hands in order to be revealed; but the hands must feel the way to where the things lead them. They are hands that require their things.

Fourth, the epistemic value of these things is transitory. They retain their research character only as long as their reference potential is not exhausted. They function after the manner of a xenotext, as Brian Rotman called it in *Signifying Nothing*.[32] A xenotext does not redeem anything definitively—it presents us with riddles. Its meaning consists in referring to further possible meanings. It lies in the essence of epistemic things that they point beyond themselves. But this promise is a peculiar one: the related expectation surrounds the object with an aura endowed with a fringe of uncertainty. It is precisely this kind of hovering expectation that drives researchers around the ledges and edges of their objects without being able to forecast precisely what will be revealed on the other side. In the end, this way of the situation being out of one's hands and yet lying within reach of interaction with an epistemic object refers back to the first point of this staccato characterization of the research situation. It also prepares the ground for the fundamental second-order category in which the sciences—as Ian Hacking impressively pointed out in *Representing and Intervening*[33]—have reference to the category of "reality" that Polanyi, to return to the above quotation, links to a thing's capacity to lead to the unprecedented. It is tied up with the ever-precarious limbo between the known, from which it aims to break off because the latter leaves something to be desired, and the unknown, upon which there is no immediate grip.

This means that "epistemological acts" are required again and again to be able to take the next step in a given research situation. It is the "unexpected impulses" that determine the "course of scientific development."[34] Such acts often happen just when the processing of the epistemic traces on paper has begun, acts where juxtapositions become possible that have no place in the material fabric of the experimental goings-on. Such scribbles, the tentative assembling and ordering of observations, are therefore an integral part of experimentation. They expand its potential and, as shown in more detail in chapter 5, they react back on the experiments. The basic problem that has to be tackled is therefore to show "how the holding of our scientific knowledge today can guide scientists to discover further knowledge, unknown and indeed inconceivable today."[35]

Phenomenology of the Concept

We can also regard the situation from another perspective by focusing on the activity of individuals concerned with these things. From the perspective of the research activity, we are dealing with acts of delegation. To build up an experiment that revolves around a particular research object and strives to explore one of the inexhaustible aspects of its thingness means to subvert the traditional subject-object relation in the sense of a direct confrontation of observer and observed. In the experiment, the act of observation is delegated to a technical arrangement suited to the task that is brought into interaction with the object of investigation.

Following Hans Blumenberg, it is then action at a distance that underlies all concept formation. Research has to be considered as a special kind of concept formation. It strives to bring what has not yet been conceptualized in everyday experience, or cannot be conceptualized because it escapes, within reach of the conceptual. For that to happen, the gestation of the experimental interaction has to occur in such a way that its result—the traces that it leaves—is not completely determined in advance. If that were the case, then the outcome would only be what we already know; we would be dealing with an act of demonstration, not one of research. A research experiment lives from a special kind of "unconceptuality," to use Blumenberg's term for this peculiar tension.[36] It must bring into the realm of the conceptual what cannot be conceptualized in advance. In any case, this is the deeper reason why concepts that reveal themselves to be productive in research exhibit fuzzy boundaries. The mathematician and philosopher of science Jean Cavaillès saw the progression of the sciences in a similar way, as "a perpetual revision of content by deepening and erasure [*rature*].... Progress is material or between singular essences, its motor the demand that each of them must be surpassed. It is not a philosophy of consciousness but a philosophy of the concept that can yield a doctrine of science. The generative necessity is not that of an activity, but of a dialectic."[37]

An experiment must have the potential to evoke unintended effects—that is, to generate unexpected results. Of course, that does not mean excluding the experimenter. On the contrary: a peculiar kind of attention is required on the part of those who carry out the experiment. Again, Polanyi comes to our aid here with his distinction between two kinds of attention: a "focal" and a "subsidiary" attention, the latter being either "subliminal" (oriented toward the interior) or "marginal" (oriented toward the exterior).[38] Focal attention fixates a thing under a defined aspect by disregarding all the others. While it allows the researcher to focus sharply on this particular point, it tends to mute the potential diversity of the object. In contrast, subsidiary attention suspends that focus in favor of an undetermined attention that covers a whole field of potentially incumbent events, none of which can be expected with certainty. It is in the nature of epistemic events that they represent boundary phenomena to begin with. This means that they would tend to escape the fully focalized senses. Under subsidiary conditions, however, we can take notice of them; consequently, they will allow the unexpected to enter. Subliminal and marginal attention helps to bring the unprecedented into the realm where it can be grasped. Claude Bernard once confided to his notebook that experimenting requires an *esprit de finesse* for feeling out things to come.[39] Therefore "the unexpected side is always more fruitful than the expected one, the consideration of natural phenomena being more instructive than the ideas we make ourselves of them."[40] However, he added by way of limitation, "the foreseen and the unforeseen stand in a reciprocal relation to each other, according to the greater or lesser developmental state of the science."[41]

Epistemicity, Experimentality

Knowledge things are therefore things that leave something to be desired. They stand for a relation to the world that is about searching for knowledge. We are concerned here with an exploratory relation that is driven by the wish to find rather than by the confirmation of what is there and given. Perhaps it is of some interest linguistically that the unit of research that usually leads to a publication is called a "finding." To quote Bernard again: "Once one has found, one reasons and applies. But where one does no longer know, one has first of all to find."[42] Experimenters—such as the physiologist Claude Bernard—are specialized in arranging situations in such a way that finding becomes possible. Scientific finding obeys neither a logic of pure accident nor a logic of pure necessity. It follows another logic, one of its own, which entails elements of chance as well as necessity, opening a third way between the stochastic rigor of the former and the deterministic rigor of the latter. One could describe such searching as a game of *eventuation*. It is an engagement with the material world that requires, first, an intimate acquaintance with the things at hand, and second (and at the same time), a distancing, the ability to let things appear strange.

It has become commonplace to use the concept of serendipity to capture the eventfulness of research. This notion, which Robert Merton introduced into the discourse on the sciences, derives from a Persian tale.[43] The notion has stuck. In an early work, *Social Theory and Social Structure*, Merton circumscribed it as follows:

> The serendipity pattern refers to the fairly common experience of observing an unanticipated, anomalous and strategic datum which becomes the occasion for developing a new theory or for extending an existing theory. . . . The datum is, first of all, unanticipated. A research directed toward the test of one hypothesis yields a fortuitous by-product, an unexpected observation, which bears upon theories not in question when the research was begun. Secondly, the observation is anomalous, surprising, either because it seems inconsistent with prevailing theory or with other established facts. . . . And thirdly, in noting that the unexpected fact must be strategic, i.e., that it must permit of implications having an impact on generalized theory, we are, of course, referring rather to what the observer brings to the datum than to the datum itself. For it obviously requires a theoretically sensitized observer to detect the universal in the particular.[44]

It is easy to see that Merton's description still orbits around the traditional terminology of theory and observation, hypothesis and test. But it nevertheless refers to the phenomenon that is at stake here, and which will reveal quite different nuances in another vocabulary.

So much for the experimental attitude of the sciences toward the world. I presume that the activities in the fields of poetry and of the arts share with the sciences the basic feature of this epistemic posture. Nonetheless, we should remember what has been said about the resilience of the *materials* in which we dwell. Thus, the counterpoint of the words against the writer's pen, the color against the paintbrush, the cells and tissues against their preparation is of a very different, often extremely different character. What unifies them all, however, is the eventfulness that confers meaning on their action. Returning to the passage from Polanyi quoted above, this is all about getting away from what is immediately at hand and obvious, circumventing what is actually given for the sake of things that are not yet realized but also cannot be anticipated, and in order to do so, making use of their "attractive power" to "manifest [themselves] in still unexpected ways." The sciences, as well as the arts, are nourished by the materials they approach and get involved with, which thus expose unprecedented aspects of them, without being able to predict the path they take and the trail they blaze by their very movements.

At stake is nothing less than an inclusive epistemology of the unprecedented. In the philosophy of science there has been a long-nurtured tradition of relegating events of discovery to the realm of intuition, and consequently psychology, and of reserving procedures of justification to epistemology.[45] Wherever one looks in this tradition, one always encounters the same pattern: separation of the irregular from the regular, the irrational from the rational, the irritating from the reassuring, the approximate from the precise. Where real research is at stake, however—not just filling the gaps, smoothing out unevenness, or specifying a magnitude—the point in the sciences, as in the arts, is to get rid of this dichotomy and to remove its apparently apodictic character.

Microhistory

In the conclusion to this chapter on knowledge and narration, I would like to tackle the historiographical consequences of this view of the research process. Experimental systems are not only to be seen as units in which epistemic events are made to happen; they can, and must, also be seen as units of historiographical narration. Here, we have to do with narrativity on the basis of historiography, to come back to Benveniste's distinction between narrativity and narrativization. The microhistory of experimental systems can be seen as a particular, epistemologically motivated mode of historical writing. It does not "follow scientists and engineers through society";[46] rather, it follows the course of epistemic things in specific experimental constellations. I have tried to explore this narrative option with a number of case studies located in the realm of the life sciences of the late nineteenth and the early twentieth centuries.[47] Such case studies face two challenges. The first consists in avoiding any appearance of teleology that haunts the retrospective view from a later state of the art as a danger—that is, exchanging a history from the past for a history toward the future. The second challenge consists in withstanding the illusion that the dynamics of the research process could simply be patched into historiography, as if historicality and narrativization could be merged and identified.

Georges Canguilhem, in particular, warned insistently against equating the object of the sciences with the object of the history of science. In his address to the Canadian Society for the History and Philosophy of Science in Montreal in 1966, "The Object of the History of Science," he pleaded for a clear distinction between the things out there, the objects of the sciences, and the object of the history of science. He summed up this distinction succinctly:

> The object in the history of sciences has nothing in common with the object of science. The scientific object, constituted by methodical discourse, is second in relation to, even though not derived from, the initial natural object which we can, playing on meanings, gladly call a pre-text. The history of sciences practices on these second, nonnatural, cultural objects but does not derive from them anymore than these second objects derived from the first. The object of historical discourse is, in effect, the historicity of scientific discourse, inasmuch as this historicity represents the carrying out of an internally normalized project, but one which is traversed by accidents, retarded or deflected by obstacles, interrupted by crises, i.e. moments of judgment and of truth.[48]

In this lecture, Canguilhem polemicizes against a historical narrative that does nothing but emulate the scientific object, thereby borrowing the respective science's own terminology. On the contrary, what is needed is a vocabulary that tries to capture the *historical* nature of the scientific development, the dynamics of a process that, following Canguilhem, is therefore "internally normalized" and at the same time "interrupted by crises." It is a dynamic that simultaneously creates the conditions of its own regulation *and* the conditions of its critical transcendence, without which this kind of historicity would not exist. From the perspective of experimental systems, that means that we have to think about the conceptual historical tools of a peculiar kind of microhistory. It stands in contrast to other, established forms of histories covering shorter time spans that have their place in the overall agenda of the history of science but are not at issue here, such as biographical narratives or histories of research institutions.[49]

The emphasis of the kind of microhistory described here is on the materiality of the epistemic process, and in particular of those aleatoric moments that happen in it and are powerful enough to point it in unforeseen directions. As described in chapter 6 using the example of Carl Correns's pea crossings, we observe here a special kind of relation between material continuity and conceptual reorientation. We could adduce many such examples, whose variants can ultimately only be understood if we focus on the subtleties of the respective experimental regimes. They are not disclosed by the published paper trail left behind by these experimental regimes. Such microhistories, therefore, require—as a rule, at least—that laboratory protocols and other unpublished sources are available that are in closer vicinity to the experimental process than the published results, and that allow us to focus on the experimental turning points in the research process as if under a magnifying glass. In the case of the botanist Correns, we can also see clearly that the laboratory protocols were not simple mnemonic devices for the later textual presentation of the results, but that their very progressions were an integral part of what was happening in the process of experimentation. That is why, for historians, they can function as a surrogate for the experimental process. With luck, what they have at their disposal is an unpublished paper trail of the experimental narrative that can help them to come to terms with the historicality—in Ricoeur's sense—of the scientific object that Canguilhem too has in his sights. Scientists work in the laboratory, historians of science in the archive. For the latter, it is understood that this leads to an important consequence: namely, to quote the microhistorian Carlo Ginzburg, "that the obstacle interfering with [historical] research in the form of lacunae or misrepresentations in the sources must become part of the account."[50]

The vocabulary needed for a historical epistemology that faces this task is not to be found fully-fledged in the annals of traditional epistemology. It remains a challenge for all those who are interested in such microhistories of the *mangle of practice*.[51] For a deeper understanding of the epistemic practices in the diverse corners and niches of the scientific universe, the multiplication of such histories remains indispensable. And there is something else to be stressed here once more. If we would like to understand how the different forms of the production of scientific knowledge connect, a bird's eye view of the theoretical products of selected sciences, such as physics, is no longer enough. What we need is a multi-perspectival view from below.

Case studies of this kind are the privileged places of narratives that are distinguished by a specific, metonymic character: on the one hand, they are very concrete, if not singular, but on the other hand they also point beyond themselves. Their form of generalization is *pars pro toto*. Exactly this distinguishes a microhistory from a mere local history. In this respect, their role in the history of science can be compared with the role of experiments in the empirical sciences. For each experiment must also be seen as a concrete, singular event. And here, too, it will only be accepted as this concrete, singular experiment if it appears at the same time as an instantiation of a phenomenon that points beyond this individual case. Otherwise, it would not be much more than a gimmick, or a *lusus naturae*. It is this above and beyond, not just in the sense of mere generalization nor of a proxy, a replacement, but in the sense of a reaching out by restriction that defines the decisive aspect of the exemplary in the first place. The exemplary in this respect[52]—if one chooses to use this concept at all here—is distinguished by a very peculiar kind of vicariousness. It does not refer to an *other*, it does not simply exemplify something more encompassing. Instead, in pointing to its own particularity it points beyond itself. It is a hypostasis.

Let me add in this context that Giovanni Levi, a prominent voice of the *microstoria* movement that originated in Italy and France in the 1970s, also insisted on comparing this kind of historical practice with "experiments rather than examples." Although he expresses skepticism about the possibility of defining a general theoretical frame for the different forms of microhistories, he nevertheless maintains that all of them are grounded by the decision to vary "the scale of observation for experimental purposes."[53] And Ginzburg, one of the paradigmatic representatives of *microstoria*, has emphatically pointed to the role of the unexpected in microhistorical explorations and insisted on the "anomalous, not the analogous" of their comparisons, in synchronic as well as diachronic perspective.[54]

Different time frames require distinct guidelines for historiographic narration. The timespan of an experimental system with its epistemic stakes does not generally go beyond the career horizon of a particular research group, and is sometimes even shorter. It corresponds roughly to what George Kubler calls a project or productivity cycle, or an "indiction period," typically between ten and twenty years.[55]

Macrohistories

To cope with the dynamic of scientific developments over longer periods of time we have to look for other reference points for directing narration, for other units whose history is to be told. The forms of narration are not indifferent to scale. We know of traditional forms of macrohistory in the realm of the sciences, such as the history of ideas[56] and the history of disciplines.[57] In their twentieth-century instantiations, they consecutively marked the rise of the historiography of science to a discipline in its own right, and they dominated the field for a long time. While the history of ideas characterized the first half, the history of disciplines rose to prominence in the second third of the century. While the former drew its resources mainly from *Geistesgeschichte*, the latter got its impulses from the sociology of science that developed in parallel.

I shall restrict myself here to a brief look at the history of disciplines. If we envision the development of the sciences over the course of the twentieth century, it is impossible not to see that the traditional disciplines are experiencing a crisis. Today, we are dealing with the emergence—and the disappearance—of fields of knowledge that are much less stringently codified and socially organized than disciplines. Their boundaries are less rigid and their constituency is more ephemeral and much more fleeting than what once constituted affiliation with a scientific discipline. Disciplines had their climax in the late nineteenth century. In the course of the twentieth century this disciplinary landscape, which was richly structured but differentiated in a largely delimited fashion, underwent a profound change.

Let me briefly outline this trend using the example of the life sciences. First, it was two hybrid sciences that took a new sounding of the terrain and challenged the laboriously but successfully stabilized boundaries between biology and physics and chemistry. These were boundaries that had already, a century earlier, been open to disposition when physiology established itself as an autonomous discipline. One of the twentieth century's newcomers was the hybrid science of biochemistry; the other was biophysics. On the one hand, both formations represented a fusion of a seemingly natural and basic scientific differentiation; and on the other hand, they assumed the character of more or less independent disciplines in their own right. Then, around the middle of the twentieth century, a formation took shape that entered history under the name of molecular biology. It presented itself as a multiple hybrid: an amalgamation of biophysical and biochemical techniques on the one hand, and genetic procedures and problems on the other. In molecular biology— and its core, molecular genetics—physics, chemistry, and biology were rearranged with respect to each other in a historically unprecedented way. And it was out of precisely this constellation that a new and previously unheard-of perception of biological specificity emerged. It revolved around the concepts of genetic "information" and genetic "program," and it made use of a language that took up and combined elements from all of these previously separated realms.[58] Self-reproduction, self-development, and self-maintenance—the three Kantian specifications of the organism—were brought under a unified paradigm that additionally made use of the terminology of cybernetics, the systems science of the time, and not only materially located these specifics in biological macromolecules, but also stamped them, by way of the information concept, with a moment of the irreducibly formal.[59]

Before long molecular biology had itself reached the status of a basic biological discipline; however, it did not enjoy a long life as a discipline.[60] Although it was the result of an enormous disciplinary hybridization, it quickly and with equal vigor surpassed itself as a discipline. The arsenal of molecular tools that emerged soon made it into an assemblage of techniques that, as such, entered into every other possible realm of the exploration of the living, and it penetrated the life sciences as a whole in a capillary fashion. The new biology that had been oriented at first toward a pure analysis of molecular structures and processes now paved the way for gene technology in its manifold applications. The prospect of a technological manipulation of the molecular basis of life not only brought a new level of analysis to the life sciences as a whole; unprecedented and further-reaching interfaces also emerged. Molecular biology was no longer an esoteric undertaking, propagated by a small flock of pure basic researchers; rather, it was soon transformed into a field in which economic and social interests began to combine with the bio-technological developmental prospects of the new approach in medicine and agriculture. The Human Genome Project was the *epistemic* expression of this new constellation,[61] and the development of a molecular biotechnology industry, with its close ties to university laboratories, was its *economic* consequence.[62]

In view of such disciplinary dynamics over the course of the past century, it is not by chance that historically oriented epistemologists reacted to these disciplinary crises. As early as the 1930s, they sought to capture the phenomenon of carving out epistemic coherence below the level of disciplines and in the interstices of the lines dividing them, and they tried to work out a conceptual frame for its description.[63] Gaston Bachelard addressed these new spaces as "cantons" or quarters, using the metaphor of science as a permanently changing city, and he associated "regional rationalisms" with them.[64] A generation later, Pierre Bourdieu introduced the notion of "field"—*champ*—in order to characterize comparatively coherent realms of social and cultural activities, including scientific practices.[65] On the one hand, he intended to make scientific forms of practice comparable with other cultural practices; on the other, and precisely through that comparison, he wanted to accentuate the aspects that were specific to scientific practice. Finally, the notion of experimental cultures, as described in their diachronic dimensions in chapter 7, can be seen as part of the tradition of such efforts, in this case with an emphasis on their temporal shape.

At this point, I would like to finally, and very briefly, touch on forms of narration that span even longer periods of time in the realm of the sciences. More precisely, the question arises as to which units could be suitable to serve as guidelines for such narratives without losing sight of scientific practice— that is, without switching to the level of that other founding discourse of the history of science, the history of ideas. From the perspective of scientific practice adopted throughout the present phenomenological essay, Michel Foucault's discursive *dispositif* could be a candidate. According to Foucault, a discourse consists of an overarching totality of practices and standards, including the convictions embodied in them, which determine together what is thinkable and sayable in this frame.[66] To articulate the dynamics of the sciences *in becoming*, "as an ensemble, at once coherent, and transformable," however, we need to be slightly more specific. Foucault himself was well aware of this when, at this critical point, he expressly referred to Georges Canguilhem.[67] The sciences unfold in the context of discourses, but they are only points of condensation within them; at the same time, however, they bear a disruptive potential. In contrast, the focus on discourse is oriented toward closure, and therefore toward an understanding of the current and the usual. What we are looking for here, however, is a narrative frame that centers around the openings and displacements over longer periods of time that have been so characteristic of the sciences.

Canguilhem has suggested writing such *longue durée* narrations as histories that follow the trajectory of *concepts* that claim a scientific status, and pursuing their vagrancies and their consecutive embodiments as well as their epistemic precipitations in different areas of investigation. He has done this extensively himself using the example of the concept of reflex during the seventeenth and eighteenth centuries.[68] We can also refer to the example of the concept of heredity, whose trajectory Staffan Müller-Wille and I have followed from the early modern period to the present.[69] Until the later eighteenth century the concept of heredity did not play a role worth mentioning in natural history. Theories of generation, whether preformistic or epigenetic, could exist without such a concept. It was only around 1800 that the concept of heredity migrated into emerging biology. It did so from the realm of law, where its traditional place was in the regulation of transmission of material property from one generation to the next. Concomitantly, the concept of generation underwent a decisive shift—from conception to cohort.[70] In the course of the nineteenth century, the idea took hold of conceptualizing generational change in terms of entities for which the organisms of a generation served as carriers. This idea entered the most diverse areas of cultural practice, such as medicine, animal and plant breeding, anthropology, and finally also evolutionary biology. Insofar as the respective practices with their related cultural techniques did not have immediate points of contact with each other, this transition led to a rather fragmented epistemic space that only coalesced into a coherent research space toward the end of the nineteenth century. This condensation of the epistemic space of heredity also led to the establishment of an epistemic entity sui generis, for which the notion of the gene became current. This concept, in turn, penetrated over the course of the twentieth century into all of the life sciences, where it was distributed among a multiplicity of experimental systems and experimental cultures, and where it also led to a series of "images of science" around this research object.[71] The first of these images was that of an "atom of life." As such, however, it remained elusive. The second was that of a material "carrier of information." Now, the atom materialized as a nucleic acid molecule, but what it meant to carry information remained largely elusive. The third image of knowledge was that of a "map." In this image the genes became nodes in a network, but the nature of that network remained poorly illuminated—and so the history continues until this day.

In this short summary, I have restricted myself to characterizing the transgressive power of the epistemic space of heredity and its materialization as genes in the realm of research—the space to which this book is devoted. This largely neglects the onto-theological and technological stories that grew up around heredity, with all their social, political, cultural and medical consequences—in other words, these materializations in the life world of the respective images of science from which, of course, the sciences cannot be uncoupled. Narratives on this level would, for instance, talk about the mechanization of the image of life, the geneticization of society, or (still more generally) the "scientification" (in the sense of the German *Verwissenschaftlichung*) of the world picture. With that, a level of generality would be reached that leads to what Jean-François Lyotard called a "master narrative."[72] Half a century ago these narrative forms were described as "ideologies."[73] However, this is a level that is not particularly illuminating from the perspective of scientific research—at least as long as one remains convinced that scientific exploration, together with a few other activities such as the arts, ultimately possesses an irresistibly subversive power. In the end, being subversive means nothing other than to have the capability of resisting such totalizations.

A narrative is a proper narrative only as long as one can imagine that it could have emerged otherwise. Narrating is thus endowed with an intrinsic

quantum of indeterminacy, of potential plurality—contrary to a strictly teleo-
logical, or ideological, structure and the usual idea of a story as the inexorable
deployment of a plot. And it is always concrete. An abstract narrative—like an
abstract experiment—is a contradiction in terms, although both the narrative
and the experiment contain a message that points beyond them. This elective
affinity in two dimensions—contingency and concreteness—is the ground on
which experimenting as narrating, and narrating as experimenting, interfere
with each other, and which ultimately allows us to think of knowledge and
narration together.

9

Thinking Wild

Science by Day, Science by Night

I begin with a paradox. Quoting an unknown authority, the French molecular biologist François Jacob once formulated it in a rather poetic fashion:

> When you look more closely at 'what scientists do,' you might be surprised to find that research actually comprises both the so-called day science and night science. Day science calls into play arguments that mesh like gears, results that have the force of certainty. . . . Conscious of its progress, proud of its past, sure of its future, day science advances in light and glory. By contrast night science wanders blind. It hesitates, stumbles, recoils, sweats, wakes with a start. Doubting everything, it is forever trying to find itself, question itself, pull itself back together. Night science is a sort of workshop of the possible, where what will become the building material of science is worked out. . . . Where thought makes its way along meandering paths and twisting lanes, most often leading nowhere. At the mercy of chance, the mind thrashes around in a labyrinth, deluged with signals, in quest of a sign, a nod, an unexpected connection.[1]

The purported anonymity in attribution—"so-called"—is not completely without significance. Jacob used it to express—and stress—the fact that we are concerned not with the experience of a solitary individual but with the collective experience of the whole scientific community. On another occasion he compared the nocturnal side of science with a *jeu des possibles*,[2] concluding that there was neither a royal path for the sciences nor an ultimate wisdom. At the end of his masterly history of heredity he cast his view into a statement and an ensuing question about the future of the life sciences: "Today the world is message, codes, and information. Tomorrow what analysis will break down our objects to reconstitute them in a new space?"[3] The nature of the sciences is such that it is exactly these questions concerning the future that cannot be answered in advance. Whereas historically sciences have developed which, given initial and boundary conditions, allow us to make predictions, the process that led to them and that will lead further on from them will nevertheless escape the principle of prediction. Our capacity of prediction appears to be limited and confined in this respect. As expounded in some detail in chapters 7 and 8, this is because the path to scientific knowledge is driven by events. One has to let them happen. We cannot derive the path from first principles and follow it like a thread of Ariadne. We can only create the conditions for the event of finding.

"In a subject such as this," Claude Lévi-Strauss states in the "Ouverture" to *The Raw and the Cooked*, the first volume—conceived as a musical panopticum—of his monumental tetralogy *Introduction to a Science of Mythology*, "scientific knowledge advances haltingly and is stimulated by contention and doubt. Unlike metaphysics it does not insist on all or nothing. . . . I shall be satisfied if it is credited with the modest achievement of having left a difficult problem in a rather less unsatisfactory state than it was before. Nor must we forget that in science there are no final truths. The scientific mind does not so much provide the right answers as ask the right questions."[4] And then, Lévi-Strauss compares his "project" explicitly with an "experiment" that "spreads like a nebula, without ever bringing together in any lasting or systematic way the sum total of the elements from which it blindly derives its substance, being confident that reality will be its guide and show it a surer road than any it might have invented. . . . It follows that as the nebula gradually spreads, its nucleus condenses and becomes more organized. Loose threads join up with one another, gaps are closed, connections are established, and something resembling order is to be seen emerging from chaos."[5]

That holds not only for anthropology but also for the natural sciences, a fact amply confirmed by Jacob's remarks. If the sciences were indeed the ultra-rationalistic endeavor that their apologists sometimes present us with, the best minds in science would probably find it uninteresting. It would no longer be a particular challenge, an intellectual adventure one would gladly get involved with. So, it is good to bow out of the current image of the sciences and, looking at the deeds and not the words of the scientists, to try and conceive an alternative image of what it means to practice science, and particularly to do research. For research is the concrete core and irreplaceable motor of all scientific knowledge, and in the center of research there is the experiment. It is the infra- and superstructures of the experiment that concern us in this treatise.

Thinking the Wild

It was Lévi-Strauss who, more than half a century ago, provided a starting point for such an attempt with his pathbreaking book on *Wild Thought*. I am, however, less concerned with characterizing wild thought in and of itself than with finding an appropriate way to think the wild kernel, the moment of the wild *within* the sciences. Nevertheless, this chapter can also be read as an homage to one of the most extraordinary thinkers of the twentieth century. Although Lévi-Strauss never wrote a systematically conceived epistemological disquisition, he provided the elements of an epistemology that would position itself beyond the distinction between the natural sciences and the humanities.

Wild Thought can be regarded as a phenomenological prelude to Lévi-Strauss's *Introduction to the Science of Mythology*, which appeared in successive volumes over the next few years. In *Wild Thought* he pleads for the appreciation of mythical thinking as a rational form of thinking in its own right: "Magical thought is not a beginning, a start, a sketch, part of an as yet unrealized whole; it forms a well-articulated system. . . . Instead, then, of opposing magic and science, we would do better to view them as parallel, as two modes of knowledge, unequal insofar as their theoretical and practical results are concerned . . . but not in the kind of mental operations on which the two draw, and which differ less in nature than as a function of the types of phenomena to which they are applied."[6] And he refers to what he calls the "Neolithic paradox": the great civilizing arts of pottery, weaving, metalworking, urban construction, jewelry making, and agriculture and livestock farming did not flow from the sciences in their modern manifestation; the sciences only made their appearance ten thousand years after the Neolithic revolution. The shapers of the latter were nevertheless driven by what Lévi-Strauss calls a "taste for knowledge"—an "appetite for objective knowledge," an "appetite for knowing and for the pleasure of knowing"[7]—for which he preferred to use the epithet "first" instead of "primitive."[8]

In the eyes of the ethnologist, mythical thinking in its orientation toward the objects of the world is distinguished from scientific thinking primarily in that the former plays out at the level of the concrete, the surface, the skin of the phenomena, while the latter operates at the level of the abstract, behind and beneath the surface of the phenomena. The former could therefore also be called phenomenological thinking. Unlike the traditional concept of phenomenology, however, this thinking is conceived not from the perspective of the subject, but from that of the world. Echoing Paul Ricoeur and with reference to

Roger Bastide, Lévi-Strauss therefore called it a "combinative, categorizing un-conscious," a "categorizing system unconnected with a thinking subject."[9] Ac-cording to him, the concrete mind, this "Science of the Concrete,"[10] is nothing like the long-abandoned antecedent of the abstract mind. He does his utmost to strip this form of thought of its illusory character, the animistic, the hylozoistic. Historically predating abstract thinking, it later entered into coexistence with it as a parallel form of disclosing the world, as a form of cognitive engagement with the world that is still indispensable to us today. Here, Lévi-Strauss meets up with Ernst Cassirer's *Philosophy of Symbolic Forms*,[11] in particular with the latter's plea for a co-presence of myth (as a form of thinking, intuiting, and liv-ing) and scientific knowledge, in their respective necessity and irreducibility to each other.

I would like, however, to take a critical step further at this point and re-gard concrete thought not only in its own domain, separated from scientific thought, but at work in the innermost core of scientific research, exactly at the spot where the exploration of what cannot be preempted scientifically happens. What needs to be shown is that concrete and abstract thinking do not stand in opposition—or even in exclusion—to each other in the practice of research, but that they presuppose each other.

Abstraction and Concretion

Gaston Bachelard, the historical epistemologist and somewhat elder contem-porary of Lévi-Strauss, also proceeded from the assumption that the concrete mind has to lay claim to a place in the center of the scientific research process. I shall present his arguments in a brief aside before returning to Lévi-Strauss and his *Wild Thought*. *Le Rationalisme appliqué*, one of Bachelard's late books on the philosophy of science from the time after the Second World War, at-tributes a complex, "*abstract-concrete* mentality" to modern physics.[12] He sees it entangled in a permanent "double action of abstraction and concretiza-tion."[13] These are not static attributions, but categories of process. Abstracting and concretizing characterizes the activity of a "field of thought"—*champ de pensée*—emerging from the "conjunction" of mathematics and experiment. "To summarize, no empty rationality thus, no incoherent empiricism—these are the two philosophical obligations that found the tight and precise syn-thesis of theory and experience, of theory and experiment in contemporary physics."[14] And Bachelard is harking back to his remarks on the relation be-tween subject and object in the process of knowledge generation in his earlier book, *The New Scientific Spirit* from 1934, when he states: "If one has to as-sure oneself of an object of scientific knowledge, one cannot confide in the

immediacy of a non-ego facing an ego."[15] He has spoken in this context of a
"strong *coupling*" of both moments, "ideas and experiments."[16]

That also means, however, that the standards the experimental proceedings actually follow are on trial in the act of the experiment itself. It is not enough for the scientific mind to "receive" impressions according to the accepted method; it must "receptionize" them,[17] to borrow Bachelard's neologism. He found even stronger words for it in his little text on "surrationalism," published in 1936: an experiment in which one does not risk one's reason is not worth carrying out.[18] We could also formulate it as follows: Scientific reasoning is bound to transcend itself, to leave behind its—always preliminary—actual state of affairs. But it is not capable of doing so in and of itself; it must deliver itself to its objects, it must try its hand on them. Edgar Wind has pointedly addressed the two "illusions" contradicting this insight: "on the one hand, the ghost of perfected science, the phantom of a logical cloud-cuckoo-land . . . on the other hand, there is the image of the human spirit which, without knowing its goal, wanders with all the confidence of the somnambulant through the sequence of stages that leads to that very cloud-cuckoo-land."[19] Echoing Bachelard, Wind states: "We thus cannot escape the conclusion that the ultimate purpose of the experiment is to test its own presupposition."[20] The researcher must enter into this feedback loop in which a form of contingency that is characteristic of the sciences plays out its role. It is a form of contingency due to that peculiar mixture of proximity and distance with which scientists must approach their materials if they are looking to recognize new aspects and to orient and reorient their thinking on these materials.

A Philosophy at Work

Above all, however, Bachelard insistently demanded that the reflection upon and contemplation of the dynamics of the sciences must not escape from such interaction, such experimental transcendence—in short, unruliness. Accordingly, he called his own philosophy of science a "philosophy at work"—*une philosophie au travail*—and contrasted it with the "philosophies of résumé," which he saw predominating in the philosophical writings of his contemporaries and which fell prey to his verdict.[21] In his critique he also included the historians of science, and he encouraged them to pay attention to the reconstruction of obscure situations in the course of science, those situations in which the abstract and the concrete, concept and object have not yet connected recursively with each other and still move in the untidy realms on the night side of knowledge.

In contrast to the epistemological tradition, Bachelard demanded that investigations on the level of a historically oriented philosophy of science

proceed in just as process-oriented a fashion as the work of the sciences referred to by—historical—epistemology: "Epistemology has to be as mobile
as science."[22] For physics this means that "every new experiment . . . exposes
the method of experimentation itself to experience"[23] (that is, to probation);
for the epistemologist it amounts to "participating in an *emergence* in order
to *understand*."[24] At stake is not a *prima philosophia*, but a "philosophy of *continuation*," and never a "philosophy of the *beginning*."[25] As an epistemologist,
too, one has to immerse oneself in this recursive process that cannot be articulated either from its beginning or from its endpoint, but only through the
peculiar form of movement that belongs to it. In this basic feature the natural
sciences do not differ from the cultural sciences and the humanities. In all
of them, things no longer happen in a Cartesian manner: "The rationalistic
cogito . . . must function like an emergence from already more or less empirically saturated holdings."[26] It sounds like an echo of this statement when
Lévi-Strauss says in passing, in the closing chapter of *Wild Thought*, alluding
to Descartes: "He who begins by installing himself in what is supposedly self-
evident to introspection never emerges from it."[27]

Why this claim to mobility for epistemology? In the end, the answer is
to be found in the specialization of the modern sciences. It was decisive for
Bachelard that with the ever closer interaction between ever more specific
forms of knowledge and the world of phenomena, the sciences split themselves into different epistemological regions. Their conceptual dynamics de-
tached them from overarching theoretical systems and images of the world
closed upon themselves, and attached them instead to concrete and very specific phenomena that were only to be realized in the experiment. To really be
understood in their phenomenotechnical constitution, the sciences could no
longer be studied from within an encompassing philosophical canon; rather,
one had to explore them in their regionally specific manifestations, in all their
details, and in the different contexts of technological application.

Bachelard's German contemporary Ernst Cassirer drew rather similar conclusions from his philosophical and historical investigations on the "problem
of knowledge," as he called it. From a contemporary theory of knowledge he
expected "a persistent, patient steeping of oneself in the work of the separate
sciences, which must not only be investigated in respect to principles, but
explained concretely, that is, in the way they conceive and handle their primary and fundamental problems."[28] For both thinkers, it was no longer obvious that, say, quantum physics and the theory of biological evolution were
organized and developed according to the same epistemological principles.
Bachelard consequently pleaded for a philosophy of the concrete scientific
act, and he saw its task as determining the multiple and different forms of

scientific thinking and researching in their respective peculiarities: how new knowledge emerged in them, how and where their results were fed into further research, and which forms of concepts were involved in the work. Bachelard did not shy away from claiming that in the limiting case, every problem, every experiment, in fact every sufficiently complex equation was in need of its own epistemology. He was convinced that it was no longer enough to proclaim general norms of scientific knowledge acquisition and to reclaim their observance. Thenceforth, the sciences claimed their own right to determine the boundaries that they required and that were in permanent flux.[29] Their only common denominator was their readiness for "rectification."[30]

Bricolage

In a well-known passage of the first chapter of *Wild Thought*, Lévi-Strauss introduces the term *bricolage*, "tinkering":

> Moreover, a form of activity still subsists among us that, on the technical plane, gives a fairly good idea of what, on the plane of speculation, might have represented what I would call a 'first science' rather than a primitive science: it is what is commonly designated by the French term *bricolage*. . . . The rule of his [the *bricoleur*'s] game is always to make do with 'what-ever is at hand'—that is to say, a set of tools and materials that is finite at each moment, as well as heterogeneous, because the composition of the set is not related to the current project, nor indeed to any given project, but is the contingent result of all the occasions that have presented themselves for renewing or enriching his stock, or for maintaining it with leftovers from earlier constructions and destructions.[31]

He contrasts it with the image of engineers who stand for modern analytical thinking, and who—at least in principle—take each of their steps under the command of a strategic plan.

Of course, we can ask whether the latter image presents engineers in the proper light. In any event, tinkering, fiddling around, improvising, and tweaking are certainly not foreign to them.[32] Above all, however, I would like to claim that the modern researcher who pursues—and promotes—a science in its empirical details cannot be a theory-guided engineer, but instead resembles a tinkerer. The appeal of a part of contemporary synthetic biology to the spirit of the engineer is thoroughly misleading.[33] At stake is not wild *thinking* as a thinking sui generis, but rather the appreciation of a moment of the *wild*, the untamable and the unplannable in the core region of scientific thought and action itself. The art of experimentation requires nothing less than that. If an experiment is conducted in such a way that it can do nothing but either corroborate or refute an assumption, which in the end makes no big difference operationally, then the experimenter remains a prisoner of the narrowness of his or her actual theoretical framework. One has to experiment in such a way that moments of the unexpected can occur. In other words, a space has to be created in which epistemic events become possible. An event in the proper sense of the word is an incident that cannot straightforwardly be deduced from what is given. Where they do research, the sciences are event-driven forms of knowledge generation. The heteroclitic composition of the materials and the instruments in the experiment favor the eventfulness of action. To quote Edgar Wind once again, this time in relation to the experimenter and the events he or she provokes, from his book on the *Experiment and Metaphysics*: "For, although we *know the meaning* of these occurrences only in terms of the preconceived system, we cannot predict their *occurrence*. What they reveal to us is the answer to a question which we have presented in logical terms, but which we cannot answer by logical means. . . . The method of his [here: the physicist] art consists in testing a purely logical conception by provoking an *entirely meta-logical act*."[34]

Working on Knowledge

In contrast to many of his analytically oriented contemporaries, Bachelard also saw this core constellation of the experiment perspicuously. His historical epistemology precludes a definite general structure of thinking or something like a "logic" of scientific knowledge, as Wind's remarks still could be interpreted to imply. Rather, he sees the generation of knowledge as an activity that constitutes and diversifies itself historically, and in which the structure of the knowledge apparatus is as deeply involved as the whole cognizant person, each of them in their varying epistemic relations. The process of knowledge generation is therefore *work* on knowledge. The "epistemological obstacle" stands in the center of its description,[35] a phenomenology of knowledge work. It is not an obstacle that imposes itself from outside, such as the over-complexity of the world or the physical and physiological limits of our senses. The epistemological obstacle arises again and again, within the process of knowledge generation itself, to the extent that new insights congeal into matters of course and thus lose their preliminary character. Bachelard formulates it as follows: "It is at the very heart of the act of cognition that, by some kind of functional necessity, sluggishness and disturbances arise."[36] These languors and turbidities are not simply preliminary delusions that have to be overcome and that can eventually be left behind as such. In a kind of structural necessity, they delay the process of knowledge acquisition on the one hand while keeping it going on the other. Immediacy has no place in this process, nor has belief in immediacy: "Reality is never 'what we might believe it to be': it is always what we ought to have thought. Empirical thought is clear *in retrospect*, when the apparatus of reason has been developed."[37] The temporal structure of knowledge acquisition is therefore that of a future past. Paul Feyerabend later formulated this insight as follows: "Theories become clear and 'reasonable' only *after* incoherent parts of them have been used for a long time."[38] And in the revised German version he added: "*Understanding always only comes after the event* and is rarely ever one of the causes of its occurrence."[39]

Ultimately, Bachelard's historical epistemology is an epistemology of error.[40] Bachelard's contemporary Hélène Metzger also focused her attention on the errors in the history of science.[41] The motif of erring is central in Georges Canguilhem's philosophy of science as well,[42] and the Italian mathematician and historian of science Federigo Enriques, to whom Bachelard referred in this context, devoted a decisive section of his essay on the *Importance of the History of Scientific Thinking* to the problem of error. Enriques wrote: "To reduce the error to the distraction of a tired mind would mean to confine it to the case of an employee who adds numbers. However, the field that has to be explored is much vaster, for what we have to do with is real intellectual work."[43] At the same time, and in a way as the flip side of error, this work is a phenomenology of "wonder."[44] The ability to discard previous knowledge as an error as well as to wonder about what delineates itself in its place are preconditions for the acquisition of scientific knowledge. André Darbon went as far as to call experimental experience a "shock" that revives the scientific spirit.[45] "[W]hat is the use of an experiment that does not rectify an error and that is just plain true and indisputable? A scientific experiment is therefore one that contradicts ordinary, everyday experience. Moreover, immediate, everyday experience always has a kind of tautological character, developing in the realm of words and definitions; what it lacks in fact is the perspective of rectified errors that in our view characterizes scientific thought."[46] However, the structure of the epistemological rupture, complementary to the epistemological obstacle, is not restricted to the transition from everyday experience to scientific experience. It continues well into the further development of the sciences. In its generalization, then, the error assumes the form of a concentration on difference—a difference that is always marked with a historical index.

In his *Contribution to a Psychoanalysis of Objective Knowledge*,[47] Bachelard devoted a whole chapter to a particular form of such an experience and its explanation that was closely connected to an image borrowed from everyday perception and that pervaded physics in the seventeenth and eighteenth centuries: the imagery of the sponge as a typical example of an epistemological obstacle. The sponge, as we know and use it in daily life, appears to us as something very obvious and empirically immediate and evident. Its structure is porous, and the fibers it consists of, while firmly connected to each other, harbor a net of cavities. For that reason, a sponge is able to absorb other materials, in particular gases and fluids, and become soaked with them while retaining its structural integrity. For René Descartes, as Bachelard points out, the sponge is the paragon of a "rarefied" body—that is, a body whose compactness is aerated and whose properties are defined by this aerated quality. "In other words," writes Bachelard, "a sponge shows us sponginess. It shows us how one particular kind of matter 'is filled' with another. This lesson in heterogeneous fullness suffices to explain everything. The metaphysics of space in Descartes is the metaphysics of the sponge."[48] This example neatly shows why Bachelard described this kind of epistemological obstacle that everyday knowledge presents to scientific knowledge as "verbal obstacle."[49] For finally, it consists of a tautology: the sponginess of the sponge. The history of medicine in the early modern period is full of explanations of this kind. But they are also multiple in physics, as Bachelard shows in his chapter on the sponge. As a prominent example he quotes René Antoine de Réaumur, who explains the compressibility of the air thus: the air is a sponge, but one even spongier than an ordinary sponge, with which the air may readily be compared—whence its extraordinary elastic properties. From this example, according to Bachelard, we can see what is meant by a "*generalized image*, which is expressed by a single word, the leitmotif of a worthless intuition."[50]

Once Again: Science, Art

The acquisition of new knowledge remains thus in the realm of a certain imponderability, of trying out, of groping, of error. In contrast to many of their contemporaries, such as Karl Popper or Hans Reichenbach, epistemologists such as Enriques, Cassirer, Bachelard, Metzger, Wind, or Cavaillès do not exclude this opaque space—or so-called context of discovery—from the domain of epistemology. Rather, they declare it epistemology's center. Nothing is forbidden here, nothing goes on at all without the opportunism of the concrete act in which knowledge is gained. If, in the words of Lévi-Strauss, the engineer at the level of technology—and the scientist at the level of theory—"always seeks to open a way through and situate himself *beyond* the constraints that make up a given state of civilization, while the *bricoleur*, willingly or by necessity, remains *on this side* of those constraints,"[51] Bachelard points to the fact that the *beyond* of the scientist can always only be gained through the *this-sidedness* of the tinkerer, thus that the lucidity of the concept is always the result of a belated process of clarification. At a practical level, the scientific objects, as objects of research, remain marked by an opaque residue that makes itself felt as a permanent challenge.

In many respects, Bachelard's conception of research thus coincides with the image that Lévi-Strauss has sketched of *artistic* creation in *Wild Thought*. For the ethnologist, art inserts itself "midway between scientific knowledge and mythical or magical thinking. For everyone knows that an artist is both something of a scientist and something of a bricoleur: with the materials and skills of a craftsman, he fashions a material object that is at the same time an object of knowledge."[52] The scientist and the *bricoleur* differ, according to Lévi-Strauss, in relation to the positions "event" and "structure" occupy in their work. Scientists induce events by means of structures—of experimental systems, in our context. Conversely, tinkerers make use of events in order to build structures.[53] But Lévi-Strauss also suspected that the situation is more complex than that suggested by this simple inversion, when elsewhere he postulates that scientific explanation "does not consist in the passage from complexity to simplicity"—that is, from the concrete to the abstract—"but in the substitution of a more intelligible complexity for one that is less so."[54] In turn, Bachelard not only ranked the science of the concrete reciprocally in line with the science of the abstract; for him, concrete thinking was the driving moment of abstraction. With that, he restored the grounding to epistemology long before Bruno Latour missed it so sorely—the grounding that Latour called for in *We Have Never Been Modern*, with these inimitable words: "When we amend the Constitution we continue to believe in the sciences, but instead of taking in their objectivity, their truth, their coldness, their extraterritoriality—qualities they have never had, except after the arbitrary withdrawal of epistemology—we retain what has always been most interesting about them: their daring, their experimentation, their uncertainty, their warmth, their incongruous blend of hybrids, their crazy ability to reconstitute the social contract."[55]

We need to track down that moment of the wild in scientific thinking. It cannot be revealed in its ready-mades, but only in their ways of making.

A Eulogy of the Fragment

The problem of the fragmentary pervades the pursuit of the experiment in all its facets. Because of its virulent ubiquity, an engagement with the fragment appears as a fitting closure to these considerations. The fragment and the fragmentary are frequently associated with discipline formation in the sciences and, more generally, with reductionism, and have negative connotations.[1] This chapter, however, is devoted to praise of the fragment.

We can view the fragmentary in relation to the scientific research process from two basic perspectives. Both refer to the dimension of time. On the one hand, there is the perspective of the past. Seen from the depths of time, the fragment is something left over, a chance residue which in the distant past once belonged to a whole, and from which, ideally, that whole can be reconstructed. On the other hand, there is the perspective of the future. Viewed from futurity, the fragment is emblematic of an open horizon, something unfinished; it is an unaccomplished piece, a construction site on whose ground the building has yet to be erected.

A Short Overview

Both past and future aspects have to be considered when talking about the fragment and the fragmentary in the context of a phenomenology of the experiment—that is, with regard to the forms of the scientific working process. I will start with a short overview of the different roles the fragmentary and fragmentation play in scientific research, in the natural sciences as well as in the historical humanities. Looking back, the perspective on the past, the *Rückfrage*, with all its epistemological intricacies,[2] stays at the center of historical research. The archive, be it the paleogeological, paleobiological, paleolithic, or historical archive in the narrower sense of written historical heritage, is *eo ipso* fragmentary. The archives only store what escaped the wear and tear of the passage of time. Every historical reconstruction must therefore necessarily proceed from what has been left. And it is precisely this incompleteness that keeps the research process in the historical sciences going, be it at the level of geological and biological evolution or at the level of human history. Paleobiologists are reminded of Georges Cuvier's famous aspiration to be able to reconstruct the complete skeleton of an animal from the fragment of a bone. From the Renaissance to our times, Classics scholars have aimed at reconstructing the transition from mythical to philosophical-scientific thinking in Greek antiquity by means of the fragments of the Pre-Socratics. Things look different in the forward-oriented research process of the natural sciences. There, the perspective on the future predominates. But the fragmentary, the incomplete also plays a decisive role, albeit in very different configurations. In particular the experiment, the central procedure of the modern natural sciences, depends on fragmentation by its very nature. Fragmentation becomes here the requisite of an entire mode of knowledge.

Scientific Research: The Laboratory

I shall begin my remarks with this second perspective. The fragmentation of reality for the sake of its analysis is the basic procedure of empirical research *tout court*. It underlies all experimentation, insofar as the experimental take on the world is inherently directed toward limitation and segmentation. At the same time, fragmentation is one of the prime difficulties of experimentation. For reality, as it is accessible to our senses and instruments, does not immediately present to the researchers the sutures along which a fruitful fragmentation might be achieved. And if it suggests fault lines and points of fracture, they usually emerge as not particularly conducive to further work. No experiment can therefore dispense with that act of "arbitrariness," as François Jacob once frankly called it,[3] without the courage to create a breach, to crack a coherent whole, to cut it into pieces. And yet, the experiment claims to point beyond its own situatedness in space and time, and with that, beyond the fragmentary character that is constitutive for it. We shall now look at the facets of the *ontic* fragmentation that underlies the experiment, and the forms in which it can be recovered *epistemically* in the research process.

To do so, it is necessary to return once again to what I have called the *infrascopic* dimension of experimentation. The basic condition is easily spelled out: the experiment can always and only be directed at a confined segment of reality. To go all out experimentally, to lay claim to totality in terms of experimentation,

would be a contradiction in terms. In the experiment, the researcher is dealing
with trimmed things, with a divided state of affairs, and usually one that escapes visibility. As shown in the early chapters of this book, ways and means
have to be found to capture them in all their detail. The kind of fragment that
is at stake here has to be prepared in such a way that, if brought into contact
and interaction with the apparatus of investigation, it leaves significant traces
behind. The contact—compare chapter 1 above—has to be such that it throws
sparks, so to speak. And the sparks have to be caught and made to last in such
a way that they can be looked at and interpreted. This is easier said than done,
however, for these traces tend to be fugacious and evanescent. The paradigmatic example of such a procedure stems from particle physics in the early
twentieth century: the visualization of subatomic particles as condensation
stripes in a cloud chamber. Charles Thomson Rees Wilson received the Nobel
Prize in 1927 for its construction. The ephemeral traces that become visible as a
condensation stripe on a screen, such as the streaks in the cloud chamber, can
be fixed durably as photographs. In this way, the trajectories can be gauged,
thus allowing inferences to be made regarding the particles that caused them.
As an alternative—and this is more and more the case today—instruments can
be constructed that register the traces of the experimental objects in a digital
form. They are then encoded directly in the language of the apparatus. The
historian of science Peter Galison has followed the development of these two
experimental traditions in microphysics closely over the twentieth century and
has described them as contrasted "image" and "logic" traditions, respectively.[4]
Image and logic, the analog and digital generation of traces, have repeatedly
challenged each other in the development of microphysics, and they have their
counterparts in the history of molecular biology as well.

    With the traces created in the experiment, we have reached the point of
utmost reduction, of maximal fragmentation. The trace is only an index of
the fragment, a fragment of the fragment. Consequently, a trace, taken as
such and in isolation, can generally no longer be interpreted. This is the point
at which the reconstitution has to begin, at which traces isolated from entire
series of experiments must be brought into relation with each other. The ontic fragmentation, the reduction in complexity that the experiment presupposes in its infrascopically oriented constitution, has to be compensated for,
embedded, and contextualized. Only through such an epistemic recontextualization can the experimental process as a whole be sustained and pointed
in directions that promise fruitful feedback. Of course, such recontextualiza-
tions can never be authoritative; they are always tentative.

    Johann Wolfgang von Goethe provided an early historical example of a
deep insight into these dialectics of experimentation—and one probably un-

expected in traditional philosophy of science. Goethe expounded the matter in a remarkable essay titled "The Experiment as Mediator of Object and Subject," an early version of which was probably written in 1792,[5] but which was only published three decades later:[6]

> The value of an experiment, whether simple or complex, is that under certain conditions, with familiar apparatus and the necessary skill, it can be at any time reproduced as long as we re-create the same situation. Rightly we stand in awe of the human mind that can bring about the necessary constellation of circumstances and that is able to craft the instruments needed for experimentation. Such are being invented daily. While we can praise a single experiment, it gains its true value only through its connection and unification with other experiments. Even to connect two experiments that are similar to each other demands more attention and vigilance than the keen observer might demand of himself. Two phenomena may be related, but not nearly as closely as we believe. Two experiments can appear to follow from one another and yet a whole series should lie between them to show the natural connections.

And Goethe concludes:

> We cannot take great enough care when making inferences based on experiments. We should not try through experiments to directly prove something or to confirm a theory. For at this pass—the transition from experience to judgment, from knowledge to application—lie in wait all our inner enemies: imaginative powers that lift us on their wings into heights while letting us believe we have our feet firmly on the ground, impatience, haste, self-satisfaction, rigidity, thought forms, preconceived opinions, lassitude, frivolity, and fickleness. This horde and all its followers lie in ambush and suddenly attack both the active observer and the quiet one who seems so well secured against all passions.[7]

A number of epistemic strategies of synthesis have been mobilized in the history of experimentation to avoid that "ambush," as Goethe calls it. In what follows I would like to briefly describe two of them, each located at a different level. The first strategy is described by Goethe himself when he states in his essay, "It is the duty of the scientist to modify every single experiment."[8] The second strategy is usually discussed in philosophy of science under the heading of model building.

As far as the first strategy, the "modification of every single experiment," is concerned, we may be dealing with very different options according to the field of experimentation and the research technologies used. Let us take a closer look at one of these forms. It comes under the heading of "independent evidence," both in philosophy of science and in the language of the actors.[9] The principle is that one tries to corroborate an experimental finding brought about by a particular research technology with another, independent technique. As a rule, what happens is more than just a corroboration—new facets of the research object are disclosed. Discrepancies in the results are no less important than coincidences. As discussed more extensively in chapter 4 above, the application of an alternative research technology simultaneously implies the creation of a new interface between the research object and the apparatus.

As an example, let us look at the determination of the structure of a nucleic acid, transfer RNA, the molecule that mediates gene expression in the cell and that first attracted the attention of those who studied protein synthesis around the middle of the 1950s. Following the history of this molecule, we see that its crystallization in an appropriate medium with subsequent X-ray diffraction analysis,[10] as conducted mainly by Alexander Rich and his group at the Massachusetts Institute of Technology over an extended period of time, played a decisive role.[11] Subsequently, the further development of nuclear magnetic resonance (NMR) spectroscopy[12] enabled the performance of a structural analysis in solution, bringing the molecule allegedly nearer to the state in which it is functionally active.[13] The two physical research technologies rest on the interaction of different constituent parts of the molecule or its atoms with the apparatus and correspondingly require different methods of preparation for their analysis. The examples taken from the history of the structure analysis of this molecule could easily be multiplied. Not least, they also comprised nearest-neighbor analysis of its building blocks, the nucleotides, which rested on physical as well as chemical and radiation-induced crosslinking (that is, the formation of covalent bonds between neighboring building blocks). In addition, they contributed to the determination of the three-dimensional shape of the molecule.[14]

The second strategy is located not in the realm of the technical conditions of the experiment, but on the level of the processing of data resulting from the traces generated in the experiment. I will concentrate here on one of its prominent facets: model building. In doing so, I will draw on the discussion of models in chapter 3. As shown there in more detail, models can accommodate not only partial data resulting from the variations of one particular series of experiments, but also those from experimental contexts that rest on different research technologies. This possibility is due not least to the fact that models perform an ontic cut with respect to the object of investigation. It is exactly this detachment that opens up the option to assemble the data produced as independent evidence at the level of model building, and in an always provisional manner. The strategy of multiple data generation in the

experiment thus achieves consistent continuation at the level of data processing in the form of model building.

Empirically derived models are combinations or textures of data in data
space. The synthetic function that they fulfill with respect to the realm of
experimentation corresponds to the synoptic function that makes the data
manageable for the beholder and allows an overview. With respect to the experiment arranged according to fragmentation, models perform a remarkable inversion. In a way, they revive the whole again, but in a problematic
fashion. From the perspective of experimental fragmentation, the model
performs a synthesis. From the perspective of the epistemic object, however,
whose model it claims to be, it always remains reduced and partial. This is
its strength, because it focuses on *one* aspect of the thing in question. It is
also its weakness, for it invites us to succumb to the illusion of dealing with a
representation of the whole.

With respect to biological research, nobody realized this entangled con-
flict situation better than Georges Canguilhem in his paper "The Role of
Analogies and Models in Biological Discovery." There he states: "Experimentation is analytical and proceeds by discriminating among the determining
conditions by varying them, all other factors being supposed unchanged. The
model method allows the comparison of entities which resist analysis. Now,
in biology, analysis is less a partition than a liberation of entities of a smaller
scale than the initial one. In this science, the use of models can legitimately
pass as being more 'natural' than elsewhere."[15]

Canguilhem points out, and rightly so, that in the life sciences experimental analysis usually leads to units of structure and function on a smaller
level—that is, to something like subordinated units. Apart from the totalizing
aspect in model comparison—and here Canguilhem presupposes the ontic
cut between epistemic object and model—the strength of models rests on two
further properties. First, they form ensembles that do not behave indiffer-
ently toward a manipulation of their own constituents. Claude Lévi-Strauss
insisted on this aspect of models again and again and emphasized that models
are "made up of elements none of which can undergo a change without ef-
fecting changes in all the other elements."[16] That systemic or "structuralist"
character—already hinted at in chapter 2—implies that models open up the
possibility of trial and triangulation in data space. The second strength is also
related to the system character of models: they can refer to data gaps. They
have the paradoxical property not only of generating an illusion of the whole,
but at the same time of subverting that illusion. In this way, the fragmentary
also claims its rights in the data space, where it takes the form of a gap. The
questions that arise from such interventions at the model level can, in turn,

be fed into the ongoing stream of the experimental production of traces. This sets a loop in motion that is based on a changeover in both directions. In the frame of a phenomenology of experimentation, this research function of models is the decisive point. Here, it is not the representational value of models that counts, but rather their operational potential. This means that the old and oft-repeated question concerning the analogical or metaphorical nature of the model moves into the background.[17] In fact, it is not even posed any longer as such.

Historical Research: The Archive

In the second part of these concluding considerations on an epistemology of the fragmentary, I would like briefly to digress on the fragment from the perspective of the archive, and in doing so, consider the past. We are dealing here with the fragmentary nature of the historical record. Both the historically oriented natural sciences and the historical cultural sciences confront us with a twofold difficulty. On the one hand, we have to reconstruct past states from historical clues and traces,[18] and on the other hand, we have to try and draw past states together into developments—that is, to extrapolate their succession and then reconstruct a historical course of events. The survival and the assessment of these clues and traces are subject to chance: on the one hand, their preservation is due to historical circumstances, and these circumstances, usually post hoc, are not causally related to what is affected by them—and what is not. On the other hand, the access to historical research in the framework of reconstruction itself depends on similar circumstances.

Let us first consider the work of the historical natural sciences.[19] Because it occupies a prominent place in this context, I will concentrate on the representative example of evolutionary biology, and on paleobiology as one of its branches.[20] Most of the early life on the planet is recorded in the sediments left by former seas. By nature, the passing on of these remnants is fragmentary and full of gaps. Individual organisms, and preferentially their more durable parts, remain preserved, either in more or less petrified form or as inclusions only under special conditions. And if they exist, the remains to be found in the layers of the sediments and analyzed are fragmentary themselves. The conditions of this conservation are manifold. With reference to the object, the factors that have to come together for such a geofact to be formed are contingent in the last analysis. Since the eighteenth century, paleobiologists have tried to reconstruct a panorama from the contingency of these fragments themselves and the likewise contingent gaps that yawn in their distribution—say, of the fauna and flora of the Precambrian, or of the hominids in the East African Olduvai Gorge.[21] One might rightly claim that sciences like this owe their very existence and their finesse to these gaps. They are powered by what the geologically embedded fossil record leaves open. Their methodologies develop around these gaps. Not only the morphology of individual organisms has to be inferred from fragments—Georges Cuvier (see above) was a master in this métier—but also the ecosystems that they formed. The image that the geological eras present to us is thus a mosaic in which most of the pieces are missing.

If we now leave the synchronous level and turn toward the diachronic, evolutionary axis, the gaps, and with them the difficulties, are potentiated. Once the horizontal layers are identified, they have to be connected vertically, and thus genealogically. The result is gaps of a second order. Which of the morphologies identified in the previous layer can rightfully be related to the ones in the next layer? The conditions of the geological record are not the same for all the sediments. If vertical gaps arise, how can we decide whether we are dealing with saltational developments, interrupted lines, or a fragmentary record? Whole generations of evolutionary biologists have quarreled about questions of "missing links," a concept dating back to the British geologist Charles Lyell.[22] Before Darwin, the so-called catastrophe theory reigned supreme. Georges Cuvier, one of its advocates, disposed of the vertical gaps by assuming that the fossils of the geological strata were bearing witness to an equal number of catastrophes, after which complete faunas and floras would have been recreated. With that assumption he could restrict his task to the reconstruction of single organisms from their remnant bones. He believed he had found the key for the closure of the gaps in his law of correlation of the organs: each fragment of a living being carried the mark of the whole to which it belonged.

Given this law, the whole could be inferred from the fragments even if it had been lost in time. Since teeth are usually particularly durable parts of skeletons, they have long played a prominent role in such reconstructions. Let us look at a passage from Cuvier's *Essay on the Theory of the Earth*, in which he even ventures a comparison of his method with mathematics:

> In short, the shape and structure of the teeth regulate the forms of the condyle, of the shoulder-blade, and of the claws, in the same manner as the equation of a curve regulates all its other properties; and, as in regard to any particular curve, all its properties may be ascertained by assuming each separate property as the foundation of a particular equation; in the same manner, a claw, a shoulder-blade, a condyle, a leg or arm bone, or any other bone separately considered, enables us to discover the description of teeth to which they have belonged; and so also reciprocally we may determine the forms of the other

bones from the teeth. Hence, commencing our investigation by a careful survey of any one bone by itself, a person who is sufficiently master of the laws of organic structure, may, as it were, reconstruct the whole animal to which that bone had belonged.[23]

So much for the gaps and their closure in paleobiology, following the overly optimistic view of one of its early proponents. A comparable situation prevails in the historical cultural sciences, such as archeology, which deal with the pre-literate history of past human cultures. Here, analogous conditions reign in relation to the record of artifacts. Its incompleteness, however, comes with its own specific qualities and peculiarities. In these closing remarks, though, I shall focus not on the archaeological archive,[24] but on a much more recent one, that of the history of science. If we restrict ourselves to the sciences of the Modern Age, the published record presents us with what is possibly the most continuous and most completely preserved archive. It is therefore not surprising that it has long remained the most authoritative record. We know, however, not just since Ludwik Fleck,[25] that although the paper trail of the sciences can give a fair overview of the state of knowledge in a particular time period, it is much less suitable for providing information on the research process and its details—that is, science in the making with all its meanderings and digressions. Fleck, for instance, categorically asserted that the "living practice" of the natural sciences differed fundamentally from its "official paper form," and that therefore the practice of the sciences with its core, the process of knowledge *acquisition*, could not be learned "from any book."[26]

This obviously opens up gaps whose occurrence appears to be inherent to the process of scientific communication. The distribution sphere of the sciences is characterized by rendering the reception of acquired knowledge as straightforward as possible. What, then, are the possibilities of bringing past knowledge acquisition processes to light, and of describing the procedures of research in an appropriate fashion? Michel Serres once compared the course of knowledge with water seeping, a "percolation" whose ramifications defy any kind of straightforwardness.[27] In tracing their convoluted paths, the historian of science is dependent on testimonies that are only transmitted through time in a fragmentary and sporadic form. It can be laboratory protocols, those primary forms of writing-up, traces of the laboratory activity, of the procedures of experimentation. For research institutions, such scribblings simply count as preliminaries for publications, and therefore usually survive only fragmentarily, if at all. And yet, they are of a particular relevance, not only for the working process in the laboratory—as described in chapter 5— but also for the historian of science interested in the intricacies of the research activity. Not only do they allow the gaps between successive publications to be closed; they also help to demarcate the logic of research from the logic of representation. The same is true for sporadically conserved laboratory arti-facts. The logic of representation itself generates gaps in our knowledge about the working process to which the represented results are owed. Historians of science become aware of them if they try to reconstruct historical experiments from the printed sources alone. Then it becomes evident that their performance required an implicit knowledge that can often be gained only by thoroughly scrutinizing the local cultural contexts.[28]

To summarize, we are confronted here with the basic problem of historiography, if only in a particular form. It can follow the sanctioned, more or less official printed traces, because even if fragmentary and marked by the sieve of history, they are stapled in the archives and survive in them. But the printed and preserved sources, as emanations and splinters of practical contexts, can also be misleading. Symbolically tainted material sources, having survived in one form or another, are, once interwoven into everyday life, the cultural historian's equivalent of the laboratory protocols and surviving laboratory devices of the historian of science. The well-known French *Annales* School,[29] and the later Italian tradition of *microstoria*[30]—itself a science of fragments in the best sense of the word—have repeatedly referred to these latter sources and highlighted their importance, although their holdings are even more fragmentary than those of the printed record. The fragment haunts historians at every turn, and yet it is at the same time their raison d'être. All their procedures orbit around sorting the splinters left by history, to multiply them through further finds, and to try over and over again to relate them to each other.

Both research in the natural sciences and cultural historical research therefore deal with fragments in a very elementary sense, and they have to try to connect them up more or less coherently. But whereas the former must split and take apart the world to understand it, the latter finds the history already in a fragmented state. The labor of fragmentation has been taken from it by the course of time. Both, however, face the task of extracting from the splinters the provisional sense of a fabric, over and over again.

Postscript

In place of a summary and conclusion, let us look at the remarkable vital whirl of the elements carved by a copper engraver (fig. 11.1). The illustration shows an engraving by Albert Flocon, a Bauhaus student in Dessau, later an engraver and teacher at the Académie des Beaux Arts in Paris, and a close friend of Gaston Bachelard. Although the image carries no title, it is easy to see that we are dealing with the traditional four elements of fire, water, air, and earth, to whose literary images Bachelard also devoted a number of books. They are engaged here in a unique, breathless vortex.

The picture is dominated by the nautilus spiral of an unbridled metamorphosis of the elements that engenders countless little spirals everywhere. It unleashes the elements into the whirl of life and sucks them up at the same time. It seizes the human beings and animals that get caught in it from outside and that drive it forward at the same time. It catches the tree with its gnarly twigs that tries to defy it, all the while being swept along, leaving little trees and twigs in its wake.

Battered by the tornado of the elements, shaken by it and at the same time resisting it, we see the contours of the clear straight lines of a construction that pushes its way out of the whirl toward the sky. The interpenetration of spiral and vertical axis, of a vital vortex and of geometry, of an all-engulfing knot and a constructive effort dominates Flocon's engraving and shows itself in the apocalyptic circle in a surrealistically tapered apotheosis. Although the black and white figures in the foreground of the image appear to be swallowed by an enormous wave, at the same time it hurls them upward, toward the sun. They are captivated in the eternal rhythm of life and death, of illumination and obfuscation, of deluge and radiant sunshine.

This gives us the keyword: it is rhythm, finally returning once again to the

FIGURE 11.1. Whirlwind of the elements
Source: Copperplate engraving by Albert Flocon, archive of the author.

shapes of time, the iterations of the experiment. In his book on the *Dialectic of Time,* Bachelard took up the concept of rhythmanalysis that goes back to the Portuguese philosopher Lúcio Alberto Pinheiro dos Santos,[1] and used it to characterize his reflections on the ground figure of all the experience of time—and every experimentation with it—as "vibrating time": "For us, the first form of time is time that vibrates."[2]

Our copperplate engraving is the epitome of a time that vibrates, the vibrating time of a gigantic metamorphosis. In its superposition of morphing events, a plate like this evokes the experience of contingency, as Bachelard remarked on the occasion of a copper engraving from Flocon's *Castles in Spain*: "I tell myself a different story from the one I told myself yesterday. Sincere contemplation is a capricious thing—pure caprice, in fact."[3] Bachelard contemplated artists' sheets with the same meticulousness he brought to the manifestations of contemporary physics.

The image prompts us to take a look again and again. It oscillates between horror and hope, tempest and dance, fire and water, earth and air. Difference lies enfolded in the act of repetition. To quote Henri Lefèbvre, the theoretician of rhythmanalysis of our time: "No rhythm without repetition in time and in space. . . . Not only does repetition not exclude differences, it also gives birth to them; it *produces* them. Sooner or later, it encounters the *event* that arrives or rather arises in relation to the sequence or series produced repetitively. In

other words: difference."[4] Or to quote Bachelard once more: "All true duration is essentially polymorphous, the real action of time requires the richness
of coincidence and the syntony of rhythmic efforts." For him, rhythm is the
"basis of all temporal effectiveness."[5] Albert Flocon's copperplate engraving
brings the rhythm of time into spatial manifestation, in the synchrony of a
representation that explodes and runs back into itself at the same time, hovering iridescently between seizure and liberation, blast and ash, construction
and destruction.

Vibrating time integrally belongs in the whirls of subatomic, microphysical
matter, the phenomena of life, individual experience, and the cultural manifestations of human activity. The experimental production of knowledge in
particular follows temporal rhythms; it is based on repetition and difference.
Knowledge, too, can be extracted from reality only in waves that simultaneously expose and engulf, split and splice.

Acknowledgments

A long period of gestation lies behind this book. Many of its topics have been presented on numerous occasions in the course of the past decade, in voice and in print, and I am thankful for all the input I received over that time. Special thanks go to the librarians of the Max Planck Institute for the History of Science in Berlin for their wonderful help with the necessary sources. Thanks also go to Karen Margolis, who carefully language-edited a draft of the text. I thank two anonymous readers for the University of Chicago Press, whose precious suggestions helped me to sharpen my arguments. Last but not least, my thanks go to Karen Darling and her crew. They competently accompanied the production of the book through its various stages from the beginning to the end.

Notes

Introduction

1. Hans-Jörg Rheinberger, *On Historicizing Epistemology. An Essay* (Stanford, CA: Stanford University Press, 2010a).

2. See, e.g., Don Ihde, *Experimental Phenomenology: Multistabilities* (Albany: State University of New York Press, 2012).

3. Hans-Jörg Rheinberger, *Toward a History of Epistemic Things: Synthesizing Proteins in the Test Tube* (Stanford, CA: Stanford University Press, 1997).

4. I have developed these notions in detail in Rheinberger, *Toward a History of Epistemic Things*.

5. Hans-Jörg Rheinberger, *An Epistemology of the Concrete: Twentieth-Century Histories of Life* (Durham, NC and London: Duke University Press, 2010b), chapters 11 and 12.

6. Ernst Cassirer, *Philosophy of Symbolic Forms*, vol. 3, *Phenomenology of Knowledge*, trans. Ralph Manheim (New Haven, CT: Yale University Press, 1953).

7. Ernst Cassirer, *The Problem of Knowledge: Philosophy, Science and History since Hegel*, trans. William H. Woglom and Charles W. Hendel (New Haven, CT: Yale University Press, 1950).

8. Cassirer, *Problem of Knowledge*, 62.

9. Examples are: Claude Gadrois, *Le Système du monde, selon les trois hypothèses, où conformément aux loix de la mechanique l'on explique dans la supposition du mouvement de la terre* (Paris: Guillaume Desprez, 1675); Carolus Linnaeus, *Systema naturae* (Leiden: Johann Wilhelm de Groot, 1735; 12th edition, Stockholm, 1766/1767); Jean Le Rond d'Alembert, *Recherches sur differens points importans du système du monde* (Paris: Antoine Boudet & J. Chardon pour David l'ainé, 1754–1756); Paul Henri Thiry, Baron d'Holbach, *Système de la nature ou des loix du monde physique & du monde moral* (London, 1770). For an example that this was also seen critically, cf. Noël André, *Théorie de la surface actuelle de la terre, ou plutôt Recherches impartiales sur le temps et l'agent de l'arrangement actuel de la surface de la terre, fondées, uniquement, sur les faits, sans système et sans hypothèse* (Paris: Société Typographique, 1806).

10. For a first orientation see Philipp Fischer et al., *Natures of Data* (Zurich and Berlin: Diaphanes, 2020).

11. Gaston Bachelard, *The New Scientific Spirit*, trans. Arthur Goldhammer (Boston: Beacon Press, 1984), 6, there misleadingly rendered as "scientific reality."

12. Michel Foucault, "Introduction," in Ludwig Binswanger, *Le Rêve et l'existence* (Brugge: Desclée de Brouwer, 1954), 9. English translation cited in Keith Alan Robinson, *Michel Foucault and the Freedom of Thought* (New York: E. Mellen Press, 2001), 1.

13. Bachelard, *New Scientific Spirit*, 13.

14. Gaston Bachelard, "Noumenon and Microphysics," *Philosophical Forum* 37 (2006): 80.

15. Gaston Bachelard, *The Formation of the Scientific Mind: A Contribution to a Psychoanalysis of Objective Knowledge*, trans. Mary McAllester Jones (Manchester: Clinamen, 2002), chapter 3.

16. Gaston Bachelard, *Le Matérialisme rationnel* (Paris: Presses Universitaires de France, 1953), 11.

17. Edmund Husserl, *Méditations cartésiennes: Introduction à la phénoménologie* (Paris: Armand Colin, 1931).

18. Bachelard, *Matérialisme rationnel*, 11. Compare also David Hyder, "Foucault, Cavaillès, and Husserl on the Historical Epistemology of the Sciences," *Perspectives on Science* 10 (2003): 107–29; see also Anton Vydra, "Gaston Bachelard and His Reactions to Phenomenology," *Continental Philosophy Review* 47 (2014): 45–58.

19. Bachelard, *Matérialisme rationnel*, 11–12.

20. Jean-Toussaint Desanti, *Phénoménologie et praxis* (Paris: Editions Sociales, 1970), 9.

21. Rheinberger, *Toward a History of Epistemic Things*; Rheinberger, *Epistemology of the Concrete*.

22. Hans Blumenberg, *Phänomenologische Schriften 1981–1988* (Berlin: Suhrkamp, 2018), 12–13, emphasis in the original.

23. Cassirer, *Problem of Knowledge*, 19.

Chapter 1

1. Gaston Bachelard, *Le Rationalisme appliqué* (Paris: Presses Universitaires de France, 1949), 56.

2. Charles Sanders Peirce, "Abduction and Induction," in *Philosophical Writings of Peirce*, ed. Justus Buchler (New York: Dover, 1955), 150–56.

3. Edgar Wind, "Some Points of Contact between History and Natural Science" (1936), in *Philosophy and History: The Ernst Cassirer Festschrift*, ed. Raymond Klibansky and Herbert James Paton (Oxford: Clarendon Press, 1936), 261.

4. In this respect, compare Terry Shinn, *Research-Technology and Cultural Change: Instrumentation, Genericity, Transversality* (Oxford: Bardwell Press, 2007).

5. See, e.g., Don Ihde, *Postphenomenology and Technoscience*, The Peking University Lectures (Albany: State University of New York Press, 2009).

6. See, e.g., Robert Brain, *The Pulse of Modernism: Physiological Aesthetics in Fin de Siècle Europe* (Spokane: University of Washington Press, 2015). More specifically, see François Dagognet, *Etienne Jules Marais: La Passion de la trace* (Paris: Hazan, 1987); Henning Schmidgen, *The Helmholtz-Curves: Tracing Lost Time* (New York: Fordham University Press, 2014a); Andreas Mayer, *The Science of Walking: Investigations into Locomotion in the Long Nineteenth Century* (Chicago: University of Chicago Press, 2020).

7. For example, early on and prominently, Bruno Latour, *Science in Action: How to Follow Scientists and Engineers Through Society* (Cambridge, MA: Harvard University Press, 1987).

8. Hans-Jörg Rheinberger, *Experiment, Differenz, Schrift* (Marburg: Basilisken-Presse, 1992).

9. Jacques Derrida, *Of Grammatology*, trans. Chakravorty Spivak (Baltimore, MD: Johns

Hopkins University Press, 1997). For a slightly different approach, taking issue with Derrida's concept of "arche-writing," see Juan Manuel Garrido Wainer, Carla Fardella, and Juan Felipe Espinosa Cristia, "Arche-writing and Data-production in Theory-oriented Scientific Practice: The Case of Free-viewing as Experimental System to Test the Temporal Correlation Hypothesis," *History and Philosophy of the Life Sciences* 43 (2021): 70.

10. Georges Didi-Huberman, *La Ressemblance par contact: Archéologie, anachronisme et modernité de l'empreinte* (Paris: Minuit, 2008).

11. Sybille Krämer, Werner Kogge, and Gernot Grube (eds.), *Spur: Spurenlesen als Orientierungstechnik und Wissenskunst* (Frankfurt am Main: Suhrkamp, 2007); see also Barbara Wittmann (ed.), *Spuren erzeugen: Zeichnen und Schreiben als Verfahren der Selbstaufzeichnung* (Zurich and Berlin: Diaphanes, 2009).

12. Charles Sanders Peirce, "Logic as Semiotic: The Theory of Signs," in *The Philosophical Writings of Peirce*, ed. Justus Buchler (New York: Dover, 1955), 98–119.

13. Rodolphe Gasché in Christine McDonald (ed.), *The Ear of the Other: Texts and Discussions with Jacques Derrida* (Lincoln, NE, and London: University of Nebraska Press, 1988), 113–14.

14. Derrida, *Of Grammatology*, 75.

15. Bachelard, *Rationalisme appliqué*, 121.

16. Jean Cavaillès, "Les Deuxièmes Cours Universitaires de Davos," in *Die II. Davoser Hochschulkurse/Les deuxièmes Cours Universitaires de Davos* (Davos: Heintz, Neu, and Zahn, 1929), 65–81, esp. 72–73.

17. Wind, "Points of Contact," 262.

18. The expression goes back to Louis Althusser, *Philosophy and the Spontaneous Philosophy of the Scientists and Other Essays*, ed. Gregory Elliot (London: Verso Books, 1989).

19. Gaston Bachelard, *The New Scientific Spirit*, trans. Arthur Goldhammer (Boston: Beacon Press, 1984), 6.

20. Robert K. Merton, "Three Fragments from a Sociologist's Notebooks: Establishing the Phenomenon, Specified Ignorance, and Strategic Research Materials," *Annual Review of Scoiology* 13 (1987): 1–28. For the distinction between data and phenomena see also James Bogen and James Woodward, "Saving the Phenomena," *The Philosophical Review* 97 (1988): 303–52; and James F. Woodward, "Data and Phenomena: A Restatement and Defense," *Synthese* 182 (2011): 165–79.

21. For a more detailed history of scintillation counters, see Hans-Jörg Rheinberger, *An Epistemology of the Concrete: Twentieth-Century Histories of Life* (Durham, NC and London: Duke University Press, 2010b), chapter 9. For a cultural history of radioactive radiation in the first half of the twentieth century, compare Maria Rentetzi, *Trafficking Materials and Gendered Experimental Practices: Radium Research in Early 20th Century Vienna* (New York: Columbia University Press, 2008), as well as Luis A. Campos, *Radium and the Secret of Life* (Chicago: University of Chicago Press, 2015).

22. Georg von Hevesy, "Historical Sketch of the Biological Application of Tracer Elements," *Cold Spring Harbor Symposia on Quantitative Biology* 13 (1948): 129–50; see also Gábor Palló, "Isotope Research Before Isotopy: George Hevesy's Early Radioactivity Research in the Hungarian context," *Dynamis* 29 (2009): 167–89.

23. For a detailed history, see Peter Galison, *Image & Logic: A Material Culture of Microphysics* (Chicago: University of Chicago Press, 1997).

24. For the history of molecular sequencing, see Miguel García-Sancho, *Biology, Computing, and the History of Molecular Sequencing: From Proteins to DNA, 1945–2000* (London: Palgrave Macmillan, 2012).

25. For its history, compare Stephen Hilgartner, *Reordering Life: Knowledge and Control in the Genomics Revolution* (Cambridge, MA: MIT Press, 2017).

26. Compare Rheinberger, *Epistemology of the Concrete*, chapter 12.

27. For the research boom after World War II compare Angela N. H. Creager, *Life Atomic: A History of Radioisotopes in Science and Medicine* (Chicago: University of Chicago Press, 2013).

28. Cf. Michael Schwab and Hans-Jörg Rheinberger, "Forming and Being Informed," in *Experimental Systems: Future Knowledge in Artistic Research*, ed. Michael Schwab (Leuven: Leuven University Press, 2013), 198–219.

29. Bruno Latour, "Drawing Things Together," in *Representation in Scientific Practice*, ed. Michael Lynch and Steve Woolgar (Cambridge, MA: MIT Press, 1990), 19–68; see also Bruno Latour, "The 'Pedofil' of Boa Vista: A Photo-philosophical Montage," *Common Knowledge* 4 (1995): 145–87.

30. Compare, in an exemplary and comprehensive fashion, Sabina Leonelli, *Data-Centric Biology: A Philosophical Study* (Chicago: University of Chicago Press, 2016).

31. See, e.g., Gabriele Gramelsberger, ed., *From Science to Computational Science: Studies in the History of Computing and Its Influence on Today's Sciences* (Zurich and Berlin: Diaphanes, 2011).

32. For the context of this development, see, e.g., Lily E. Kay, *Who Wrote the Book of Life? A History of the Genetic Code* (Stanford, CA: Stanford University Press, 2000).

33. Sabina Leonelli, "Documenting the Emergence of Bio-ontologies; or, Why Researching Bioinformatics Requires HPSSB," *History and Philosophy of the Life Sciences* 32 (2010): 105–26.

34. Pnina Abir-Am, "From Biochemistry to Molecular Biology: DNA and the Acculturated Journey of the Critic of Science Erwin Chargaff," *History and Philosophy of the Life Sciences* 2 (1980): 3–60.

35. On the relation between epistemic and technical things, see Hans-Jörg Rheinberger, *Toward a History of Epistemic Things: Synthesizing Proteins in the Test Tube* (Stanford, CA: Stanford University Press, 1997), esp. chapter 1.

36. As to the topic of intersection, compare Rheinberger, *Epistemology of the Concrete*, chapter 11.

Chapter 2

1. For more detail, see Hans-Jörg Rheinberger, *An Epistemology of the Concrete: Twentieth-Century Histories of Life* (Durham, NC and London: Duke University Press, 2010b), chapter 12.

2. The literature on scientific model building is voluminous. Relevant for the present context are three more recently edited volumes: Axel Gelfert, ed., "Model-Based Representation in Scientific Practice," *Studies in History and Philosophy of Science* 42, no. 2, special issue (2011); Bas van Fraassen and Isabelle Peschard, eds., *The Experimental Side of Modeling* (Minneapolis: University of Minnesota Press, 2018); Bernd Mahr, *Schriften zur Modellforschung* (Paderborn: Mentis Verlag, 2022).

3. The language use of the sciences themselves is not unambiguous in this respect. Biochemical "model systems," for instance, would have to be addressed as preparations. Alternatively, following a suggestion by Christoph Hoffmann, we could use the concept of model as the superordinate category under which preparations would have to be subsumed as a special kind of model (Christoph Hoffmann, "Probe, Konservat, Modell—Präparat?" in *Präparat Bergsturz*, ed. Katharina Ammann and Priska Gisler [Lucerne, Poschiavo and Chur: Edizioni Periferia and Bündner Kunstmuseum, 2012], 61–69); concerning problems of demarcation, see also Maria

Keil, "Verwischte Grenzen zwischen Präparat und Modell: Epistemische Mustererkennung in medizinhistorischen Sammlungen," *Medizinhistorisches Journal* 55 (2020): 75–85. Here I will retain the chosen distinction between preparations and models.

4. For three-dimensional models in the sciences see, e.g., Soraya de Chadarevian and Nick Hopwood, eds., *Models: The Third Dimension of Science* (Stanford, CA: Stanford University Press, 2004).

5. See, e.g., Sergio Sismondo, "Models, Simulations, and Their Objects," *Science in Context* 12 (1999): 247–60; Margaret Morrison, *Reconstructing Reality: Models, Mathematics, and Simulations* (Oxford and New York: Oxford University Press, 2015); see esp. Franck Varenne, *From Models to Simulations* (New York: Routledge, 2019), and Ariana Borrelli and Janina Wellmann, eds., "Computer Simulations," *NTM Zeitschrift für Geschichte der Wissenschaften, Technik und Medizin* 27, no. 4, special issue (2019).

6. Ludwik Fleck, *Genesis and Development of a Scientific Fact*, trans. Frederick Bradley and Thaddeus J. Trenn (Chicago: University of Chicago Press, 1979).

7. Hans-Jörg Rheinberger, *Toward a History of Epistemic Things: Synthesizing Proteins in the Test Tube* (Stanford, CA: Stanford University Press, 1997).

8. Cf. esp. Bruno Latour and Steve Woolgar, *Laboratory Life: The Construction of Scientific Facts*, 2nd edition (Princeton, NJ: Princeton University Press, 1986).

9. Claude Lévi-Strauss, *Wild Thought*, trans. Jeffrey Mehlman and John Leavitt (Chicago: University of Chicago Press, 2021), 28.

10. Bernd Mahr, "Das Mögliche im Modell und die Vermeidung der Fiktion," in *Science & Fiction: Über Gedankenexperimente in Wissenschaft, Philosophie und Literatur*, ed. Thomas Macho and Annette Wunschel (Frankfurt am Main: Fischer, 2004), 161–82.

11. Richard Levins, "The Strategy of Model Building in Population Biology," *American Scientist* 54 (1966): 421–31, esp. 421–22.

12. Cf. Morrison, *Reconstructing Reality*, 20; see also Mary S. Morgan, *The World in the Model: How Economists Work and Think* (Cambridge: Cambridge University Press, 2012).

13. Evelyn Fox Keller, "Models Of and Models For: Theory and Practice in Contemporary Biology," *Philosophy of Science* 67, supplement (2000): 72–86.

14. Alfred Tarski, *Introduction to Logic and the Methodology of Deductive Sciences* (Oxford: Oxford University Press, 1941).

15. See, e.g., Reinhard Wendler, *Das Modell zwischen Kunst und Wissenschaft* (Munich: Fink, 2013); Albena Yaneva, "Scaling Up and Down: Extraction Trials in Architectural Design," *Social Studies of Science* 35 (2005): 867–94.

16. The literature on model organisms is voluminous, to mention only a few: the pioneering article of Richard M. Burian, "How the Choice of Experimental Organism Matters: Epistemological Reflections on an Aspect of Experimental Practice," *Journal of the History of Biology* 26 (1993): 351–67; Angela N. H. Creager, *The Life of a Virus: Tobacco Mosaic Virus as an Experimental Model, 1930–1965* (Chicago: University of Chicago Press, 2002); Karen Rader, *Making Mice: Standardizing Animals for American Biomedical Research* (Princeton, NJ: Princeton University Press, 2004); Gabriel Gachelin, ed., *Les Organismes modèles dans la recherche médicale* (Paris: Presses Universitaires de France, 2006), and esp. Jean Gayon, "Les Organismes modèles en biologie et en médecine," in *Les Organismes modèles dans la recherche médicale*, ed. Gabriel Gachelin (Paris: Presses Universitaires de France, 2006), 9–43; Jim Endersby, *A Guinea Pig's History of Biology* (Cambridge, MA: Harvard University Press, 2009); Rachel A. Ankeny, "Historiographic Reflections on Model Organisms; or, How the Mureaucracy May Be Limiting Our Understanding of

Contemporary Genetics and Genomics," *History and Philosophy of the Life Sciences* 32 (2010): 91–104; Rachel A. Ankeny and Sabina Leonelli, "What's So Special about Model Organisms?" *Studies in History and Philosophy of Science Part A* 42 (2011): 313–23; Christian Reiß, *Der Axolotl: Ein Labortier im Heimaquarium, 1864–1914* (Göttingen: Wallstein, 2020); Rachel A. Ankeny and Sabina Leonelli, *Model Organisms* (Cambridge: Cambridge University Press, 2021).

17. Cf. Arno Levy and Adrian Currie, "Model Organisms Are Not Models," *British Journal for the Philosophy of Science* 66 (2015): 327–48.

18. Rheinberger, *Epistemology of the Concrete*, chapter 11.

19. Cf. Hans-Jörg Rheinberger, "A History of Protein Synthesis and Ribosome Research," in *Protein Synthesis and Ribosome Structure: Translating the Genome*, ed. Knud H. Nierhaus and Daniel N. Wilson (Weinheim: Wiley-VCH, 2004), 1–51

20. As to the mediating function of models, compare the classic essay collection of Mary S. Morgan and Margaret Morrison, eds., *Models as Mediators: Perspectives on Natural and Social Science* (Cambridge: Cambridge University Press, 1999).

21. For the concept of "immutable mobiles," compare chapter 1 above with reference to Bruno Latour, "Drawing Things Together," in *Representation in Scientific Practice*, ed. Michael Lynch and Steve Woolgar (Cambridge, MA: MIT Press, 1990).

22. Cf. Margaret Morrison, "Models as Autonomous Agents," in *Models as Mediators: Perspectives on Natural and Social Science*, ed. Mary S. Morgan and Margaret Morrison (Cambridge: Cambridge University Press, 1999), 38–65; see also Bernd Mahr and Reinhard Wendler, *Modelle als Akteure: Fallstudien*, KIT-Report 156, Technische Universität Berlin, 2009, https://www.flp .tu-berlin.de/fileadmin/fg53/KIT-Reports/r156.pdf.

23. Hans-Jörg Rheinberger, "Cytoplasmic Particles in Brussels (Jean Brachet, Hubert Chantrenne, Raymond Jeener) and at Rockefeller (Albert Claude), 1935–1955," *History and Philosophy of the Life Sciences* 19 (1997): 47–67.

24. James D. Watson, *Molecular Biology of the Gene* (New York: W. A. Benjamin, 1965).

25. Cf. Peter Godfrey-Smith, "The Strategy of Model-based Science," *Biology and Philosophy* 21 (2006): 725–40; Roman Frigg, "Models and Fiction," *Synthese* 172 (2010): 251–68; Mauricio Suarez, ed., *Fictions in Science: Philosophical Essays on Modeling and Idealization* (London: Routledge, 2009).

26. See Ian Hacking, *Representing and Intervening: Introductory Topics in the Philosophy of Natural Science* (Cambridge: Cambridge University Press, 1983).

27. Alain Badiou, *Le Concept de modèle* (Paris: Librairie Arthème Fayard, 2007), 58.

28. Peter Machamer, Lindley Darden, and Carl F. Craver, "Thinking about Mechanisms," *Philosophy of Science* 67 (2000): 1–25; Carl Craver and Lindley Darden, eds., "Mechanisms in Biology," *Studies in History and Philosophy of Biological and Biomedical Sciences* 36, no. 2, special issue (2005). See also William Bechtel, *Discovering Cell Mechanisms: The Creation of Modern Cell Biology* (Cambridge: Cambridge University Press, 2006); Mathias Grote, *Membranes to Molecular Machines: Active Matter and the Remaking of Life* (Chicago: University of Chicago Press, 2019); and for a critical view, Daniel J. Nicholson, "Is the Cell Really a Machine?" *Journal of Theoretical Biology* 477 (2019): 108–26.

29. Gottlob Frege, "On Sense and Reference," in *Translations from the Philosophical Writings of Gottlob Frege*, 3rd edition, ed. Peter Geach and Max Black (Oxford: Blackwell, 1980), 56–78.

30. Levins, "Strategy of Model Building," 423; see also Sara Green, "When One Model Is Not Enough: Combining Epistemic Tools in Systems Biology," *Studies in History and Philosophy of Biological and Biomedical Sciences* 44 (2013): 170–80.

31. Georges Canguilhem, "The Role of Analogies and Models in Biological Discovery," in *Scientific Change*, ed. Alistair C. Crombie (London: Heinemann, 1963), 514.

32. Compare P. Nigel T. Unwin and C. Taddei, "Packing of Ribosomes in Crystals from the Lizard *Lacerta sicula*," *Journal of Molecular Biology* 114 (1977): 491–506.

33. Conversation at the symposium on the fiftieth anniversary of the Max Planck Institute for Molecular Genetics in Berlin, December 2014.

34. Jean Baudrillard, *Simulations* (New York: Semiotext[e], 1983), 31–32.

35. Gabriele Gramelsberger, ed., *From Science to Computational Science: Studies in the History of Computing and Its Influence on Today's Sciences* (Zurich and Berlin: Diaphanes, 2011).

36. Morrison, *Reconstructing Reality*, 218; Evelyn Fox Keller, "Models, Simulations, and Computer Experiments," in *The Philosophy of Scientific Experimentation*, ed. Hans Radder (Pittsburgh: University of Pittsburgh Press, 2003), 205.

37. Varenne, *From Models to Simulations*, 9.

38. Varenne, *From Models to Simulations*, 10.

Chapter 3

1. Literature on visualization in the sciences has grown enormously. For an early and introductory collection, see Michael Lynch and Steve Woolgar, eds., *Representation in Scientific Practice* (Cambridge, MA: MIT Press, 1990), and Catelijne Coopmans, Janet Vertesi, Michael Lynch, and Steve Woolgar, eds., *Representation in Scientific Practice Revisited* (Cambridge, MA: MIT Press, 2013). See also Michael Lynch, "The Production of Scientific Images: Vision and Re-vision in the History, Philosophy, and Sociology of Science," *Communication & Cognition* 31 (1998): 213–28; Norton Wise, "Making Visible," *Isis* 97 (2006): 75–82; Klaus Sachs-Hombach, ed., *Bildtheorien: Anthropologische und kulturelle Grundlagen des Visualistic Turn* (Frankfurt am Main: Suhrkamp, 2009). For an overview on newer literature, see Regula Valérie Burri and Joseph Dumit, "Social Studies of Scientific Imaging and Visualization," in *The Handbook of Science and Technology Studies*, ed. Edward J. Hackett, Olga Amsterdamska, Michael Lynch, and Judy Wajcman (Cambridge, MA: MIT Press, 2008), 297–317.

2. Johann Wolfgang von Goethe, "The Experiment as Mediator of Object and Subject," *Context* 24 (Fall 2010): 19–23.

3. Cf. Friedrich Steinle, "Das Nächste ans Nächste reihen: Goethe, Newton und das Experiment," *Philosophia naturalis: Archiv für Naturphilosophie und die philosophischen Grenzgebiete der exakten Wissenschaften und Wissenschaftsgeschichte* 39 (2002): 141–72; Olaf Müller, *Mehr Licht: Goethe mit Newton im Streit um die Farben* (Frankfurt am Main: Fischer, 2015).

4. Cf. Christoph Hoffmann, *Unter Beobachtung: Naturforschung im Zeitalter der Sinnesapparate* (Göttingen: Wallstein, 2006).

5. On maps, see Rasmus Grønfeldt Winther, *When Maps Become the World* (Chicago: University of Chicago Press, 2020).

6. On the history of mapping in genetics, see Hans-Jörg Rheinberger and Jean-Paul Gaudillière, eds., *The Mapping Cultures of Twentieth Century Genetics*, Vol. 1, *Classical Genetic Research and Its Legacy*, and Jean-Paul Gaudillière and Hans-Jörg Rheinberger, eds., *The Mapping Cultures of Twentieth Century Genetics*, Vol. 2, *From Molecular Genetics to Genomics* (London: Routledge, 2004).

7. Emil Du Bois-Reymond, "Über die Lebenskraft: Aus der Vorrede zu den Untersuchungen über tierische Elektrizität (1848)," in *Vorträge über Philosophie und Gesellschaft*, ed. Siegfried Wollgast (Hamburg: Felix Meiner, 1974), 3–24, esp. 4.

8. For a more extensive discussion, see Hans-Jörg Rheinberger, *An Epistemology of the Concrete: Twentieth-Century Histories of Life* (Durham, NC and London: Duke University Press, 2010b), chapter 11.

9. Ilana Löwy, ed., "Microscopic Slides: Reassessing a Neglected Historical Resource," *History and Philosophy of the Life Sciences* 35, no. 3, special issue (2013).

10. Cf. Rheinberger, *Epistemology of the Concrete*, chapter 12.

11. Hanna Lucia Worliczek, *Wege zu einer molekularisierten Bildgebung: Eine Geschichte der Immunfluoreszenzmikroskopie als visuelles Erkenntnisinstrument der modernen Zellbiologie (1959–1980)*, PhD diss., University of Vienna, 2020.

12. Since the year 2000, for example, there has been a biennial international meeting on "Diagrammatic Reasoning and Inference" regularly documented by Springer. See also Frederick Stjernfeldt, *Diagrammatology: An Investigation on the Borderlines of Phenomenology, Ontology, and Semiotics* (Dordrecht: Springer, 2007). For the context of cultural, media, and image studies in Germany, see Matthias Bauer and Christoph Ernst, *Diagrammatik: Einführung in ein kultur- und medienwissenschaftliches Forschungsfeld* (Bielefeld: Transcript, 2010); Eckart Lutz et al., eds., *Diagramm und Text: Diagrammatische Strukturen und die Dynamisierung von Wissen und Erfahrung* (Wiesbaden: Reihert, 2014); *Diagrammatik-Reader: Grundlegende Texte aus Theorie und Geschichte* (Berlin: De Gruyter, 2016); Sybille Krämer, *Figuration, Anschauung, Erkenntnis: Grundlinien einer Diagrammatologie* (Berlin: Suhrkamp, 2016); Astrit Schmid-Burkhardt, *Die Kunst der Diagrammatik: Perspektiven eines neuen bildwissenschaftlichen Paradigmas* (Bielefeld: Transcript, 2017).

13. James D. Watson, *Molecular Biology of the Gene* (New York: W. A. Benjamin, 1965); see also Reinhard Wendler, "Visuelles Denken in James Watsons 'The Molecular Biology of the Gene'," in *Long Lost Friends: Zu den Wechselbeziehungen zwischen Design-, Medien- und Wissenschaftsforschung*, ed. Christoph Windgätter and Claudia Mareis (Zurich: Diaphanes, 2013), 23–37.

14. For the genesis of the representation of chemical processes by formulae in the nineteenth century see Ursula Klein, *Experiments, Models, Paper Tools: Cultures of Organic Chemistry in the Nineteenth Century* (Stanford, CA: Stanford University Press, 2002).

15. On the disdain of biochemistry by first generation molecular biologists, see, e.g., Mahlon Hoagland, *Toward the Habit of Truth: A Life in Science* (New York: W. W. Norton, 1990); see also Pnina Abir-Am, "The Politics of Macromolecules: Molecular Biologists, Biochemists, and Rhetoric," *Osiris* 7 (1992): 164–91.

16. Emil Fischer, "Einfluss der Configuration auf die Wirkung der Enzyme," *Berichte der Deutschen Chemischen Gesellschaft* 27 (1894): 2985–93; see also Frieder W. Lichtenthaler, "Hundert Jahre Schlüssel-Schloss-Prinzip: Was führte Emil Fischer zu dieser Analogie?" *Angewandte Chemie* 106 (1994): 2456–67; and Rebecca Mertens, *The Construction of Analogy-Based Research Programs: The Lock-and-Key Analogy in 20th Century Biochemistry* (Bielefeld: Transcript, 2019).

17. Paul Ehrlich, "On Immunity, with Special Reference to Cell Life," *Proceedings of the Royal Society London* 66 (1900): 424–48.

18. Heinrich Hertz, *The Principles of Mechanics Presented in a New Form*, trans. D. E. Jones and J. T. Walley (London: Macmillan and Company, 1899).

19. Ernst Mach, *The Science of Mechanics*, trans. Thomas J. McCormack (Chicago: The Open Court Publishing Company, 1892); Ernst Mach, *Contributions to the Analysis of Sensations*, trans. C. M. Williams (Chicago: The Open Court Publishing Company, 1897).

Chapter 4

1. See François Delaporte, *Figures de la médecine* (Paris: Les Editions du Cerf, 2009); Uwe Wirth and Veronika Sellier, eds., *Impfen, Pfropfen, Transplantieren* (Berlin: Kadmos, 2011).

2. Isabelle Stengers, ed., *D'Une science à l'autre: Des concepts nomades* (Paris: Le Seuil, 1987); see also Mieke Bal, *Travelling Concepts in the Humanities: A Rough Guide* (Toronto: University of Toronto Press, 2002); and Peter Howlett and Mary S. Morgan, eds., *How Well Do Facts Travel? The Dissemination of Reliable Knowledge* (Cambridge: Cambridge University Press, 2010).

3. Alexis Lepère, *Pratique raisonnée de la taille du pêcher*, 2nd edition (Paris: Bouchard-Huzard, 1846), 13.

4. Sarah Jansen, "An American Insect in Imperial Germany: Visibility and Control in Making the Phylloxera in Germany, 1870–1914," *Science in Context* 13 (2000): 31–70.

5. More recent investigations show that even an exchange of relevant biomolecules can take place between the two systems. Cf. Lei Yang et al., "m^{5}C Methylation Guides Systemic Transport of Messenger RNA Over Graft Junctions in Plants," *Current Biology* 29 (2019): 1–12.

6. Cf. Anne-Marie Moulin, *L'Aventure de la vaccination* (Paris: Fayard, 1996); Christine Holmberg, *The Politics of Vaccination: A Global History* (Manchester: Manchester University Press, 2017). For France, see for example Bruno Latour, *The Pasteurization of France* (Cambridge, MA: Harvard University Press, 1993).

7. Thomas Schlich, *The Origins of Organ Transplantation: Surgery and Laboratory Science, 1880–1930* (Rochester, NY: University of Rochester Press, 2010); concerning the cultural historical context, cf. François Delaporte, Bernard Devauchelle, and Emmanuel Fournier, eds., *Transplanter: Une approche transdisciplinaire: Art, médecine, histoire et biologie* (Paris: Hermann, 2015).

8. On the importance of hybridization for the history of genetics, see Staffan Müller-Wille and Hans-Jörg Rheinberger, *A Cultural History of Heredity* (Chicago: University of Chicago Press, 2012). For the use of the concept in postcolonial discourse, see Néstor García Canclini, *Hybrid Cultures* (Minneapolis: University of Minnesota Press, 1990). On the anthropological discourse in general, see Bruno Latour, *We Have Never Been Modern* (Cambridge, MA: Harvard University Press, 1993); see also Yvonne Spielmann, *Hybridkultur* (Frankfurt am Main: Suhrkamp, 2009).

9. Hans Stubbe, *Kurze Geschichte der Genetik* (Jena: Fischer, 1963); Leslie C. Dunn, *A Short History of Genetics: The Development of Some of the Main Lines of Thought, 1864–1939* (New York: McGraw-Hill, 1965); Robert Olby, *Origins of Mendelism* (New York: Schocken Books, 1967).

10. Edna Suarez Díaz, "Variation, Differential Reproduction and Oscillation: The Evolution of Nucleic Acid Hybridization," *History and Philosophy of the Life Sciences* 35 (2013): 39–44.

11. Isabelle Stengers, "La Propagation des concepts," in *D'Une science à l'autre: Des concepts nomades*, ed. Isabelle Stengers (Paris: Seuil, 1987), 9–26.

12. Cf., e.g., Jacques Ascher and Jean-Pierre Jouet, *La Greffe, entre biologie et psychanalyse* (Paris: Presses Universitaires de France, 2004).

13. Jacques Derrida, "Signature Event Context," in *Limited Inc.*, trans. Samuel Weber (Evanston, IL: Northwestern University Press, 1988), 1–23.

14. Jacques Derrida, *Dissemination*, trans. Barbara Johnson (Chicago: University of Chicago Press, 1981), 287–366.

15. Cf. Uwe Wirth, "Zitieren Pfropfen Exzerpieren," in *Kreativität des Findens—Figurationen des Zitats*, ed. Martin Roussel (Munich: Fink, 2012), 79–98.

16. Derrida, *Dissemination*, 355.

17. See Hans-Jörg Rheinberger, *The Hand of the Engraver: Albert Flocon Meets Gaston Bachelard*, trans. Kate Sturge (Albany: State University of New York Press, 2018).

18. Gaston Bachelard, *Water and Dreams: An Essay on the Imagination of Matter*, trans. Edith Farrell (Dallas: The Pegasus Foundation, 1983), 1.

19. Bachelard, *Water and Dreams*, 1, emphasis in original.

20. Bachelard, *Water and Dreams*, 3.

21. Bachelard, *Water and Dreams*, 10.

22. Bachelard, *Water and Dreams*, 3, emphasis in original.

23. Bachelard, *Water and Dreams*, 10. Pierre Alechinsky, a member of CoBrA since 1949, dwelt on Bachelard's attempt in this initial chapter of the book to "get to the bottom of matter"—*de peser une matière* (p. 26). Pierre Alechinsky, "Abstraction faite," *Cobra* 10 (Fall 1951): 4–8. Alechinsky talks here about the "unpredictable" of the "stubborn presence" of matter (p. 6).

24. Pierre Bourdieu, *The Logic of Practice*, trans. Richard Nice (Cambridge: Polity Press, 1990); Pierre Bourdieu and Loïc J. D. Wacquant, *An Invitation to Reflexive Sociology* (Chicago: University of Chicago Press, 1992); and Loïc Wacquant and Aksu Akçaoglu, "Pratique et pouvoir symbolique chez Bourdieu vu de Berkeley," *Revue de l'Institut de Sociologie (ULB)* 86 (2016): 35–60.

25. Gaston Bachelard, *Le Rationalisme appliqué* (Paris: Presses Universitaires de France, 1949), 133.

26. Compare also Hans-Jörg Rheinberger, *An Epistemology of the Concrete: Twentieth-Century Histories of Life* (Durham, NC and London: Duke University Press, 2010b), chapter 13.

27. Claude Lévi-Strauss, *Wild Thought*, trans. Jeffrey Mehlman and John Leavitt (Chicago: University of Chicago Press, 2021), 21 (the English translation renders "heteroclite" as "heterogeneous").

28. For the concept of "iteration," see Derrida, "Signature Event Context," 8–9, 18.

29. Fritz Heider, *Ding und Medium* (Berlin: Kadmos, 2005), 24–25.

30. Cf. Hans-Jörg Rheinberger, *Toward a History of Epistemic Things: Synthesizing Proteins in the Test Tube* (Stanford, CA: Stanford University Press, 1997), chapter 6.

31. Rheinberger, *Epistemology of the Concrete*, chapter 6. On *Ephestia* as model organism, see chapter 3 of the present book.

32. Rheinberger, *Epistemology of the Concrete*, chapter 11.

33. The historical literature on this case is relatively abundant. For an overview compare Michel Morange, *The Black Box of Biology: A History of the Molecular Revolution* (Cambridge, MA: Harvard University Press, 2020), chapter 14; for a more detailed account see Lily E. Kay, *Who Wrote the Book of Life? A History of the Genetic Code* (Stanford, CA: Stanford University Press, 2000), chapter 5; see also Laurent Loison and Michel Morange, eds., *L'Invention de la régulation génétique: Les Nobel 1965 (Jacob, Lwoff, Monod) et le modèle de l'opéron dans l'histoire de la biologie* (Paris: Editions rue d'Ulm, 2017).

34. Cf. Rheinberger, *Epistemology of the Concrete*, chapter 10.

35. See Müller-Wille and Rheinberger, *Cultural History of Heredity*.

Chapter 5

1. The present work only deals with protocols in the laboratory. For the protocol in its many other forms as a cultural technique, see Michael Niehaus and Hans-Walter Schmidt-Hannisa, eds., *Das Protokoll. Kulturelle Funktionen einer Textsorte* (Frankfurt am Main: Peter Lang, 2005).

2. From the perspective of science studies, cf. Bruno Latour and Steve Woolgar, *Laboratory*

Life: The Construction of Scientific Facts, 2nd edition (Princeton, NJ: Princeton University Press, 1986). A more recent overview from the perspective of history of science is given by Frederic L. Holmes, Jürgen Renn, and Hans-Jörg Rheinberger, eds., *Reworking the Bench: Research Notebooks in the History of Science* (Dordrecht: Kluwer, 2003); see also Christoph Hoffmann, *Die Arbeit der Wissenschaften* (Zurich and Berlin: Diaphanes, 2013).

3. Among the exceptions are Christoph Hoffmann and Peter Berz, eds., *Über Schall: Ernst Machs und Peter Salchers Geschossfotografien* (Göttingen: Wallstein, 2001), and Christoph Hoffmann, *Schreiben im Forschen* (Tübingen: Mohr Siebeck, 2018).

4. In the realm of the life sciences, Holmes is perhaps the most prolific writer in this respect, with books on scientists from the eighteenth century to the twentieth (Antoine Laurent de Lavoisier, Claude Bernard, Hans Krebs, Matthew Meselson, and Seymour Benzer). See, e.g., Frederic L. Holmes, *Hans Krebs: Architect of Intermediary Metabolism* (Oxford: Oxford University Press, 1993); and more generally, Frederic L. Holmes, *Investigative Pathways: Patterns and Stages in the Careers of Experimental Scientists* (New Haven, CT: Yale University Press, 2004).

5. Gerd Grasshoff, Robert Casties, and Kärin Nickelsen, *Zur Theorie des Experiments: Untersuchungen am Beispiel der Entdeckung des Harnstoffzyklus* (Bern: Bern Studies in the History and Philosophy of Science, 2001); Kärin Nickelsen and Gerd Grasshoff, "Concepts from the Bench: Krebs and the Urea Cycle," in *Going Amiss in Experimental Research*, ed. Giora Hon et al. (Boston: Springer, 2009), 91–117.

6. See, e.g., Claude Bernard, *Cahier de notes 1850–1860*, complete edition of the "Cahier rouge", ed. Mirko Dražen Grmek (Paris: Gallimard, 1965); Gerd Grasshoff and Kärin Nickelsen, *Dokumente zur Entdeckung des Harnstoffzyklus*, Vol. 1, *Laborbuch Hans Krebs und Erstpublikationen* (Bern: Bern Studies in the History and Philosophy of Science, 2001).

7. Peter B. Medawar, *The Art of the Soluble* (London: Methuen, 1967). See also Frederic L. Holmes, "Wissenschaftliches Schreiben und wissenschaftliches Entdecken," in *Schreiben als Kulturtechnik*, ed. Sandro Zanetti (Frankfurt am Main: Suhrkamp, 2012), 412–40.

8. Cf. Lorraine Daston, "Taking Notes," *Isis* 95 (2004): 443–48. A well-known example from the eighteenth century are the *Sudelbücher* by the physicist and natural philosopher Georg Christoph Lichtenberg, in *Schriften und Briefe*, ed. Wolfgang Promies, vols. 1 and 2 (Munich: Hanser, 1967). Cf. Rüdiger Campe, "Kritzeleien im Sudelbuch: Zu Lichtenbergs Schreibverfahren," in *Über Kritzeln*, ed. Christian Criesen et al. (Zurich: Diaphanes, 2012), 165–88; Rüdiger Campe, "'Unsere kleinen blinden Fertigkeiten': Zur Entstehung des Wissens und zum Verfahren des Schreibens in Lichtenbergs Sudelbüchern," *Lichtenberg-Jahrbuch* 2011 (2012): 7–32; see also Elisabetta Mengaldo, *Zwischen Naturlehre und Rhetorik. Kleine Formen des Wissens in Lichtenbergs Sudelbüchern* (Göttingen: Wallstein, 2021).

9. Hoffmann and Berz, eds., *Über Schall*, 47–141.

10. Hoffmann and Berz, eds., *Über Schall*, 85, 133–34.

11. Bernard, *Cahier de notes 1850–1860*.

12. Mirko Grmek, "Introduction," in Claude Bernard, *Cahier de notes 1850–1860*, ed. Mirko Grmek (Paris: Gallimard, 1965), 17.

13. Grmek, "Introduction," 18.

14. Grmek, "Introduction," 13, 21.

15. Claude Bernard, *An Introduction to the Study of Experimental Medicine*, trans. Henry Copley Greene (New York: Macmillan, 1927).

16. Claude Bernard, *Lectures on the Phenomena of Life Common to Animals and Plants*, trans. Hebbel E. Hoff, Roger Guillemin, and Lucienne Guillemin (Springfield: Charles C. Thomas, 1974).

17. Hans Blumenberg, "Paradigmen zu einer Metaphorologie," *Archiv für Begriffsgeschichte* 6 (1960): 7–142, 301–5.

18. Joseph Sambrook, Edward F. Fritsch, and Tom Maniatis, *Molecular Cloning: A Laboratory Manual* (Cold Spring Harbor, NY: Cold Spring Harbor Laboratory Press, 1982).

19. Cf. Angela N. H. Creager, "Recipes for Recombining DNA: A History of Molecular Cloning: A Laboratory Manual," in *Learning by the Book: Manuals and Handbooks in the History of Knowledge*, ed. Angela Creager et al., *BJHS Themes* 5, special issue (2020): 225–43.

20. A good overview on the literature is to be found in Gerald L. Geison and Frederic L. Holmes, eds., *Research Schools: Historical Reappraisals* (*Osiris* 8) (Chicago: University of Chicago Press, 1993).

21. For a theory of notation in general, see Nelson Goodman, *Languages of Art: An Approach to a Theory of Symbols* (Indianapolis, IN and Cambridge: Hackett Publishing Company, 1976), chapter 4; the anthropological *locus classicus* concerning lists is Jack Goody, "What's In a List?" in *The Domestication of the Savage Mind* (Cambridge: Cambridge University Press, 1977), 74–111; see also James Delbourgo and Staffan Müller-Wille, eds., Focus Section "Listmania," *Isis* 103 (2012): 710–52, and Staffan Müller-Wille and Isabelle Charmantier, "Lists as Research Technologies," *Isis* 103 (2012): 743–52.

22. Emil Du Bois-Reymond, "Über die Lebenskraft: Aus der Vorrede zu den Untersuchungen über tierische Elektrizität (1848)," in *Vorträge über Philosophie und Gesellschaft*, ed. Siegfried Wollgast (Hamburg: Felix Meiner, 1974), 8.

23. Du Bois-Reymond, *Über die Lebenskraft*, 6.

24. Cf. Bruno Latour, "Drawing Things Together," in *Representation in Scientific Practice*, ed. Michael Lynch and Steve Woolgar (Cambridge, MA: MIT Press, 1990); see also Hoffmann and Bertz, *Über Schall*.

25. Bruno Latour gives a beautiful example of such reversibility in "The 'Pedofil' of Boa Vista: A Photo-philosophical Montage," *Common Knowledge* 4 (1995): 145–87.

26. An expression found in the Berlin physiologist Johannes Müller's "Jahresbericht über die Fortschritte der anatomisch-physiologischen Wissenschaften im Jahre 1833," *Archiv für Anatomie, Physiologie und wissenschaftliche Medicin* (1834): 3–4; see also Hans-Jörg Rheinberger, "From the 'Originary Phenomenon' to the 'System of Pelagic Fishery': Johannes Müller (1801–1858) and the Relation Between Physiology and Philosophy," in *From Physico-Theology to Bio-Technology: Essays in the Social and Cultural History of Biosciences. A Festschrift for Mikulas Teich*, ed. Kurt Bayertz and Roy Porter (Amsterdam: Rodopi, 1998), 133–52.

27. Cf. Friedrich Steinle, *Exploratory Experiments: Ampère, Faraday, and the Origins of Electrodynamics* (Pittsburgh: University of Pittsburgh Press, 2016).

28. Cf. Thomas Macho and Annette Wunschel, eds., *Science & Fiction: Über Gedankenexperimente in Wissenschaft, Philosophie und Literatur* (Frankfurt am Main: Fischer, 2004).

29. Michel Foucault, *The Archaeology of Knowledge* & *The Discourse on Language*, trans. A. M. Sheridan Smith (New York: Pantheon Books, 1972), 140.

30. See, e.g., Christian Licoppe, *La Formation de la pratique scientifique: Le discours de l'expérience en France et en Angleterre (1630–1820)* (Paris: Editions La Découverte, 1996).

31. In the case of Germany and Austria, an overview is given by Heinz Fortak and Paul Schlaak, eds., *100 Jahre Deutsche Meteorologische Gesellschaft in Berlin 1884–1984* (Berlin: Selbstverlag, 1984). See also Stefan Emeis and Cornelia Lüdecke, eds., *From Beaufort to Bjerknes and Beyond: Critical Perspectives on Observing, Analyzing, and Predicting Weather and Climate* (Augsburg: Erwin Rauner Verlag, 2005).

32. See e.g., Christoph Hoffmann, *Unter Beobachtung: Naturforschung im Zeitalter der Sinnesapparate* (Göttingen: Wallstein, 2006).

33. Cf. Bernd Gausemeier, "Pedigrees of Madness: The Study of Heredity in Nineteenth and Early Twentieth Century Psychiatry," *History and Philosophy of the Life Sciences* 36 (2015): 467–83; Theodore M. Porter, *Genetics in the Madhouse: The Unknown History of Human Heredity* (Princeton, NJ: Princeton University Press, 2018); Soraya de Chadarevian, *Heredity Under the Microscope: Chromosomes and the Study of the Human Genome* (Chicago: University of Chicago Press, 2021).

34. For an early paper on the topic see Stephen Hilgartner, "Biomolecular Databases: New Communication Regimes for Biology?" *Science Communication* 17 (1997): 506–22; for the present state of the art see Sabina Leonelli, *Data-Centric Biology: A Philosophical Study* (Chicago: University of Chicago Press, 2016); Bruno Strasser, *Collecting Experiments: Making Big Data Biology* (Chicago: University of Chicago Press, 2019).

35. An example is given by Hans Hofmann in Philipp Fischer et al., *Natures of Data* (Zurich and Berlin: Diaphanes, 2020), 125–48; see also Michael F. McGovern, "Genes Go Digital: *Mendelian Inheritance in Man* and the Genealogy of Electronic Publishing in Biomedicine," *The British Journal for the History of Science* 54, no. 2 (2021): 1–19.

36. For drawing and writing as procedures of research compare Christoph Hoffmann, ed., *Daten sichern. Schreiben und Zeichnen als Verfahren der Aufzeichnung* (Zurich and Berlin: Diaphanes, 2008).

37. Sabine Ammon, *Epistemologie des Entwerfens: Operative Artefakte und Visualisierungstechniken in Architektur und Ingenieurwesen*, Habilitationsschrift, Technische Universität Darmstadt, 2018; Barbara Wittmann, ed., *Werkzeuge des Entwerfens* (Zurich and Berlin: Diaphanes, 2018).

38. Friedrich Kittler, *Discourse Networks 1800/1900*, trans. Michael Metteer with Chris Cullens (Stanford, CA: Stanford University Press, 1990). The German title says "Aufschreibesysteme."

Chapter 6

1. For the latter, see Henning Schmidgen, ed., *Lebendige Zeit: Wissenskulturen im Werden* (Berlin: Kadmos, 2005); Henning Schmidgen, *Hirn und Zeit: Die Geschichte eines Experiments* (Berlin: Matthes & Seitz, 2014b).

2. Helga Nowotny, *Time: The Modern and Postmodern Experience*, trans. Neville Plaice (Cambridge: Polity Press, 1996), 6.

3. Cf. Hayden White, *The Content of the Form: Narrative Discourse and Historical Representation* (Baltimore, MD and London: Johns Hopkins University Press, 1987).

4. Krzysztof Pomian, *L'Ordre du temps* (Paris: Gallimard, 1984).

5. Cf. Reinhart Kosellek, *Sediments of Time: On Possible Histories*, trans. Sean Franzel and Stefan-Ludwig Hoffmann (Stanford, CA: Stanford University Press, 2018).

6. George Kubler, *The Shape of Time. Remarks on the History of Things* (New Haven, CT and London: Yale University Press, 1962).

7. For an instructive recent analysis, see Önur Erdur, *Die epistemologischen Jahre: Philosophie und Biologie in Frankreich, 1960–1980* (Zurich: Chronos, 2018).

8. Kubler, *Shape of Time*, 122.

9. On this point, see Edgar Zilsel, "Geschichte und Biologie, Überlieferung und Vererbung" (1931), in *Wissenschaft und Weltauffassung: Aufsätze 1929–1933*, ed. Gerald Mozetic (Vienna: Böhlau, 1998), 101–44.

10. For Kubler's perspective on things, compare Esther Pasztory, *Thinking with Things: Toward a New Vision of Art* (Austin: University of Texas Press, 2005); see also Sarah Maupeu, Kerstin Schankweiler, and Stefanie Stallschus, eds., *Im Maschenwerk der Kunstgeschichte: Eine Revision von George Kublers "The Shape of Time"* (Berlin: Kadmos, 2014).

11. Peter Coveney and Roger Highfield, *Frontiers of Complexity* (London: Faber & Faber, 1995), 39.

12. Ernst Cassirer, *The Logic of the Cultural Sciences: Five Studies*, trans. Steve G. Lofts (New Haven, CT: Yale University Press, 2000), 98.

13. Kubler, *Shape of Time*, vii.

14. Ernst Cassirer, *Philosophy of Symbolic Forms*, Vol. 1, *Language*; Vol. 2, *Mythical Thought*; Vol. 3, *The Phenomenology of Knowledge*, trans. Ralph Manheim (New Haven, CT: Yale University Press, 1953–1957).

15. Kubler, *Shape of Time*, 10.

16. Kubler, 11.

17. Kubler, 10.

18. Edgar Zilsel, *Die Geniereligion: Ein kritischer Versuch über das moderne Persönlichkeitsideal, mit einer historischen Begründung* (Frankfurt am Main: Suhrkamp, 1990).

19. Kubler, *Shape of Time*, 125.

20. Kubler, 16.

21. Thomas S. Kuhn, *The Structure of Scientific Revolutions*, 2nd edition, enlarged (Chicago: University of Chicago Press, 1970), 172.

22. Kuhn, 171.

23. Kubler, *Shape of Time*, esp. chapters 2 and 4.

24. Kubler, 37–38.

25. Ludwik Fleck, *Genesis and Development of a Scientific Fact*, trans. Frederick Bradley and Thaddeus J. Trenn (Chicago: University of Chicago Press, 1979), 14–15.

26. Fleck, 96.

27. Fleck, 86, emphasis in the original.

28. Fleck, 93, passim.

29. Kubler, *Shape of Time*, 35–39, 109–12.

30. For the trajectory of an in vitro system for the investigation of protein biosynthesis, see Hans-Jörg Rheinberger, *Toward a History of Epistemic Things: Synthesizing Proteins in the Test Tube* (Stanford, CA: Stanford University Press, 1997); for other examples see also Hans-Jörg Rheinberger, *An Epistemology of the Concrete: Twentieth-Century Histories of Life* (Durham, NC and London: Duke University Press, 2010b), chapters 4 to 7.

31. Cf. Rheinberger, *Toward a History of Epistemic Things*; Lily E. Kay, *Who Wrote the Book of Life? A History of the Genetic Code* (Stanford, CA: Stanford University Press, 2000); Christina Brandt, *Metapher und Experiment: Von der Virusforschung zum genetischen Code* (Göttingen: Wallstein, 2004).

32. For greater detail, see Rheinberger, *Epistemology of the Concrete*, chapter 4.

33. See Evelyn Fox Keller, *The Century of the Gene* (Cambridge, MA: Harvard University Press, 2000).

34. My colleague Staffan Müller-Wille and I have devoted a book-length essay to this topic: Hans-Jörg Rheinberger and Staffan Müller-Wille, *The Gene: From Genetics to Postgenomics* (Chicago: University of Chicago Press, 2017).

35. See also Peter Howlett and Mary S. Morgan, eds., *How Well Do Facts Travel? The Dissemination of Reliable Knowledge* (Cambridge: Cambridge University Press, 2010); for the migration of data, compare Sabina Leonelli, *Data-Centric Biology: A Philosophical Study* (Chicago: University of Chicago Press, 2016).

36. To mention only the molecular biological research technologies, almost all of them have received their own monographical account. For example, Paul Rabinow, *Making PCR: A Story of Biotechnology* (Chicago: University of Chicago Press, 1996); Nicolas Rasmussen, *Picture Control: The Electron Microscope and the Transformation of Biology in America, 1940–1960* (Stanford, CA: Stanford University Press, 1997); and Bruno Strasser, *La Fabrique d'une nouvelle science: La biologie moléculaire à l'âge atomique* (Florence: Olschki, 2006); Angela N. H. Creager, *Life Atomic: A History of Radioisotopes in Science and Medicine* (Chicago: University of Chicago Press, 2013); Miguel García-Sancho, *Biology, Computing, and the History of Molecular Sequencing: From Proteins to DNA, 1945–2000* (London: Palgrave Macmillan, 2012). For research technologies in general, see Bernward Joerges and Terry Shinn, *Instrumentation Between Science, State, and Industry* (Dordrecht: Kluwer, 2001).

37. For light microscopy see, as a more recent example, Jutta Schickore, *The Microscope and the Eye: A History of Reflections, 1740–1870* (Chicago: University of Chicago Press, 2007); for scanning tunnel microscopy cf. Jochen Hennig, *Bildpraxis: Visuelle Strategien in der frühen Nanotechnologie* (Bielefeld: Transkript, 2011); for electron microscopy, see Eric Lettkemann, *Stabile Interdisziplinarität: Eine Biografie der Elektronenmikroskopie aus historisch-soziologischer Perspektive* (Baden-Baden: Nomos, 2016); see also Falk Müller, *Jenseits des Lichts: Siemens, AEG und die Anfänge der Elektronenmikroskopie in Deutschland* (Göttingen: Wallstein, 2021).

38. See Bruno Latour, *Science in Action: How to Follow Scientists and Engineers Through Society* (Cambridge, MA: Harvard University Press, 1987).

39. Robert K. Merton, "Three Fragments from a Sociologist's Notebooks: Establishing the Phenomenon, Specified Ignorance, and Strategic Research Materials," *Annual Review of Scoiology* 13 (1987): 1–28.

40. Phineas W. Whiting, "Rearing Meal Moths and Parasitic Wasps for Experimental Purposes," *The Journal of Heredity* 12 (1921): 255–61, esp. 258.

41. Alfred Kühn and Karl Henke, *Genetische und entwicklungsphysiologische Untersuchungen an der Mehlmotte* Ephestia kühniella Zeller, *I–VI*, Abhandlungen der Gesellschaft der Wissenschaften zu Göttingen, Mathematisch-Physikalische Klasse, Neue Folge, vol. 15 (Berlin: Weidmannsche Buchhandlung, 1929), 2.

42. Albrecht Hase, "Insekten," in *Methodik der wissenschaftlichen Biologie*, vol. 2, ed. Tibor Péterfi (Berlin: Springer, 1928), 265–89.

43. Cf. Robert E. Kohler, *Lords of the Fly: Drosophila Genetics and the Experimental Life* (Chicago: University of Chicago Press, 1994).

44. For more detail, see Rheinberger, *Epistemology of the Concrete*, chapter 6.

45. On the technique of transplantation, see chapter 4.

46. Claude Bernard, *Philosophie: Manuscrit inédit*, ed. Jacques Chevalier (Paris: Editions Hatier-Boivin, 1954), 14.

47. As to the different stages of this trajectory, compare the collation of original papers by Karen-Beth G. Scholthof, John G. Shaw, and Milton Zaitlin, eds., *Tobacco Mosaic Virus: One Hundred Years of Contributions to Virology* (St. Paul, MN: APS Press, 1999).

48. Isabelle Stengers, ed., *D'Une science à l'autre: Des concepts nomades* (Paris: Seuil, 1987); Howlett and Morgan, *How Well Do Facts Travel?*

49. Cf. James Griesemer and Grant Yamashita, "Zeitmanagement bei Modellsystemen: Drei Beispiele aus der Evolutionsbiologie," in *Lebendige Zeit*, ed. Henning Schmidgen (Berlin: Kadmos, 2005), 213–41.

50. Kubler, *Shape of Time*, 36.

51. Kubler, 36.

52. Werner Heisenberg, *Wandlungen in den Grundlagen der Naturwissenschaft. Zwei Vorträge* (Leipzig: S. Hirzel, 1935), 7, 16–17.

53. Mahlon Hoagland, *Toward the Habit of Truth: A Life in Science* (New York: W. W. Norton, 1990), xx.

54. Kubler, *Shape of Time*, 91.

55. Kubler, 32.

Chapter 7

1. For the concept of experimental culture, see Ursula Klein, "Paper Tools in Experimental Cultures," *Studies in History and Philosophy of Science Part A* 32 (2001): 265–302. For the concept of knowledge cultures, or epistemic cultures, see Karin Knorr Cetina, *Epistemic Cultures: How the Sciences Make Knowledge* (Cambridge, MA: Harvard University Press, 1999).

2. The *locus classicus* is Alistair Crombie, *Styles of Scientific Thinking in the European Tradition: The History of Argument and Explanation Especially in the Mathematical and Biomedical Sciences and Arts* (London: Gerald Duckworth & Company, 1995).

3. Ian Hacking, "'Style' for Historians and Philosophers," *Studies in History and Philosophy of Science Part A* 23 (1992): 1–20; for a new twist in using this concept see Caroline Ehrhardt, "E uno plures? The Many Galois Theories (1832–1900)," in *Cultures without Culturalism: The Making of Scientific Knowledge*, ed. Karine Chemla and Evelyn Fox Keller (Durham, NC and London: Duke University Press, 2017), 327–51.

4. John V. Pickstone, *Ways of Knowing: A New History of Science, Technology and Medicine* (Manchester: Manchester University Press, 2000); John V. Pickstone, "A Brief Introduction to Ways of Knowing and Ways of Working," *History of Science* 49 (2011): 235–45.

5. Cf. Clifford Geertz, *The Interpretation of Cultures* (New York: Basic Books, 1973); see also Ernst Cassirer, *Philosophy of Symbolic Forms*, Vols. 1–3, trans. Ralph Manheim (New Haven, CT: Yale University Press, 1953–1957).

6. Thomas S. Kuhn, *The Structure of Scientific Revolutions*, 2nd edition, enlarged (Chicago: University of Chicago Press, 1970); see also Gary Gutting, *Paradigms and Revolutions: Applications and Appraisals of Thomas Kuhn's Philosophy of Science* (Notre Dame, IN: University of Notre Dame Press, 1980); Michael Fischer and Paul Hoyningen-Huene, *Paradigmen: Facetten einer Begriffskarriere* (Frankfurt am Main and New York: Peter Lang, 1997); George A. Reisch, *The Politics of Paradigms: Thomas S. Kuhn, James B. Conant, and the Cold War "Struggle for Men's Minds"* (Albany: State University of New York Press, 2019).

7. Compare also Angela N. H. Creager, "Paradigms and Exemplars Meet Biomedicine," in *Kuhn's* Structure of Scientific Revolutions *at Fifty: Reflections on a Science Classic*, ed. Lorraine Daston and Robert Richards (Chicago: University of Chicago Press, 2016), 151–66.

8. George M. Gould, *Illustrated Dictionary of Medicine, Biology, and Allied Sciences* (Philadelphia: P. Blakiston Son & Co., 1894), 623b; see also Kijan Espahangizi, "From Topos to Oikos:

The Standardization of Glass Containers as Epistemic Boundaries in Modern Laboratory Research (1850–1900)," *Science in Context* 28 (2015): 397–425.

9. The classic reference in this respect is Robert E. Kohler, *From Medical Chemistry to Biochemistry: The Making of a Biomedical Discipline* (Cambridge: Cambridge University Press, 1982).

10. Eduard Buchner, "Alkoholische Gährung ohne Hefezellen," *Berichte der Deutschen Chemischen Gesellschaft* 30 (1897): 117–24. We shall not deal with the prioritization struggle around this event here. Cf. Robert E. Kohler, "The Background to Eduard Buchner's Discovery of Cellfree Fermentation," *Journal of the History of Biology* 4 (1971): 35–61; Milton Wainwright, "Early History of Microbiology," *Advances in Applied Microbiology* 52 (2003): 333–55.

11. Herbert C. Friedmann, "From Friedrich Wöhler's Urine to Eduard Buchner's Alcohol," in *New Beer in an Old Bottle: Eduard Buchner and the Growth of Biochemical Knowledge*, ed. Athel Cornish-Bowden (València: Universitat de València, 1997), 108.

12. Claude Bernard, *Lectures on the Phenomena of Life Common to Animals and Plants*, trans. Hebbel E. Hoff, Roger Guillemin and Lucienne Guillemein (Springfield: Charles C. Thomas, 1974), 145–72.

13. Bernard, 83.

14. Mirko D. Grmek, *Raisonnement expérimental et recherches toxicologiques chez Claude Bernard* (Geneva and Paris: Librairie Droz, 1973); Frederic L. Holmes, *Claude Bernard and Animal Chemistry* (Cambridge, MA: Harvard University Press, 1974).

15. Auguste Fernbach and Louis Hubert, "De l'Influence des phosphates et de quelques autres matières minérales sur la diastase protéolytique du malt," *Comptes Rendus des Séances de l'Académie des Sciences* 131 (1900): 293–95; Søren P. L. Sørensen, "Enzymstudien. II. Mitteilung. Über die Messung und die Bedeutung der Wasserstoffionenkonzentration bei enzymatischen Prozessen," *Biochemische Zeitschrift* 21 (1909): 131–304; Leonor Michaelis and Heinrich Davidsohn, "Die Abhängigkeit der Trypsinwirkung von der Wasserstoffionenkonzentration," *Biochemische Zeitschrift* 36 (1911): 280–90; Leonor Michaelis and Maud Menten, "Die Kinetik der Invertinwirkung," *Biochemische Zeitschrift* 49 (1913): 333–69.

16. For a history of biochemistry from a disciplinary perspective, see Kohler, *From Medical Chemistry to Biochemistry*; Joseph S. Fruton, *Molecules and Life: Historical Essays on the Interplay of Chemistry and Biology* (New York: Wiley, 1972).

17. Friedmann, "From Wöhler's Urine to Buchner's Alcohol," 110.

18. Richard Willstätter and Margarete Rohdewald, "Über enzymatische Systeme der Zucker-Umwandlung im Muskel," *Enzymologia* 8 (1940): 20.

19. Willstätter and Rohdewald, 51–52.

20. The voluminous literature on the history of bacteriology cannot be reviewed here. Cf. Thomas D. Brock, *The Emergence of Bacterial Genetics* (New York: Cold Spring Harbor Laboratory Press, 1990) and Silvia Berger, *Bakterien in Krieg und Frieden: Eine Geschichte der medizinischen Bakteriologie in Deutschland 1890–1933* (Göttingen: Wallstein, 2009); see also William C. Summers, "From Culture as Organism to Organism as Cell: Historical Origin of Bacterial Genetics," *Journal of the History of Biology* 24 (1991): 171–90; Andrew Mendelsohn, " 'Like All that Lives': Biology, Medicine and Bacteria in the Age of Pasteur and Koch," *History and Philosophy of the Life Sciences* 24 (2002): 3–36.

21. See Frederick B. Churchill, ed., "Special Section on Protozoology," *Journal of the History of Biology* 22 (1989): 185–323; see also Hans-Jörg Rheinberger, *An Epistemology of the Concrete: Twentieth-Century Histories of Life* (Durham, NC and London: Duke University Press, 2010b), chapter 5.

22. Ross G. Harrison, "The Outgrowth of the Nerve Fiber as a Mode of Protoplasmic Movement," *Journal of Experimental Zoology* 9 (1910): 787–846; see also Jane Maienschein, *Transforming Traditions in American Biology, 1880–1915* (Baltimore, MD and London: Johns Hopkins University Press, 1991).

23. Alexis Carrel and Montrose T. Burrow, "Cultivation of Tissue In Vitro and Its Technique," *Journal of Experimental Medicine* 13 (1911): 387–96; see also Jan A. Witkowski, "Alexis Carrel and the Mysticism of Tissue Culture," *Medical History* 23 (1979): 279–96.

24. Hannah Landecker, "Building 'a New Type of Body in Which to Grow a Cell': Tissue Culture at the Rockefeller Institute, 1910–1914," in *Creating a Tradition of Biomedical Research: Contributions to the History of the Rockefeller University*, ed. Darwin H. Stapleton, 151–74 (New York: The Rockefeller University Press, 2004).

25. Hannah Landecker, *Culturing Life: How Cells Became Technologies* (Cambridge, MA: Harvard University Press, 2007).

26. Petra Werner, *Otto Warburg: Von der Zellphysiologie zur Krebsforschung* (Berlin: Verlag Neues Leben, 1988), 131.

27. Otto Warburg, "Über die katalytischen Wirkungen der lebendigen Substanz," in *Über die katalytischen Wirkungen der lebendigen Substanz: Arbeiten aus dem Kaiser Wilhelm-Institut für Biologie Berlin-Dahlem* (Berlin: Julius Springer, 1928), 1.

28. Alfred Kühn to Ernst Caspari, December 27, 1946, in Reinhard Mocek, ed., *Alfred Kühn (1885 bis 1968)—Lebensbilder in Briefen* (Rangsdorf: Basilisken-Presse, 2020), 182.

29. Alfred Kühn, "Grenzprobleme zwischen Vererbungsforschung und Chemie," *Berichte der Deutschen Chemischen Gesellschaft* 71 (1938): 107.

30. Jean-Paul Gaudillière and Ilana Löwy, eds., *The Invisible Industrialist* (London: Palgrave Macmillan, 1998); Karen Rader, *Making Mice: Standardizing Animals for American Biomedical Research* (Princeton, NJ: Princeton University Press, 2004); Jim Endersby, *A Guinea Pig's History of Biology* (Cambridge, MA: Harvard University Press, 2009).

31. For examples and a methodological discussion, see William Bechtel, *Discovering Complexity* (Princeton, NJ: Princeton University Press, 1993); and Bechtel, *Discovering Cell Mechanisms: The Creation of Modern Cell Biology* (Cambridge: Cambridge University Press, 2006).

32. Compare Ilana Löwy, "Variances in Meaning in Discovery Accounts: The Case of Contemporary Biology," *Historical Studies in the Physical and Biological Sciences* 21 (1990): 87–121; Hans-Jörg Rheinberger, "Vom Mikrosom zum Ribosom: 'Strategien' der 'Repräsentation' 1935–1955," in *Die Experimentalisierung des Lebens: Experimentalsysteme in den biologischen Wissenschaften 1850/1950*, ed. Hans-Jörg Rheinberger and Michael Hagner (Berlin: Akademie Verlag, 1993), 162–87.

33. Cf. Hans-Jörg Rheinberger, "Cytoplasmic Particles in Brussels (Jean Brachet, Hubert Chantrenne, Raymond Jeener) and at Rockefeller (Albert Claude), 1935–1955," *History and Philosophy of the Life Sciences* 19 (1997): 47–67.

34. For the usage of this concept, see George E. Palade, "Intracellular Distribution of Acid Phosphatase in Rat Liver Cells," *Archives of Biochemistry* 30 (1951): 144.

35. For details see Angela N. H. Creager, *Life Atomic: A History of Radioisotopes in Science and Medicine* (Chicago: University of Chicago Press, 2013).

36. Cf. Rheinberger, *Epistemology of the Concrete*, chapter 9.

37. Michel Morange, *The Black Box of Biology: A History of the Molecular Revolution* (Cambridge, MA: Harvard University Press, 2020); Hans-Jörg Rheinberger, "Kurze Geschichte der Molekularbiologie," in *Geschichte der Biologie*, ed. Ilse Jahn (Jena: Gustav Fischer, 1998), 642–63.

38. See, e.g., Pnina Abir-Am, "From Biochemistry to Molecular Biology: DNA and the Accul-turated Journey of the Critic of Science Erwin Chargaff," *History and Philosophy of the Life Sci-ences* 2 (1980): 3–60; Seymour Cohen, "The Biochemical Origins of Molecular Biology," *Trends in Biochemical Sciences* 9 (1984): 334–36; Joseph S. Fruton, *Proteins, Enzymes, Genes: The Inter-play of Chemistry and Biology* (New Haven, CT: Yale University Press, 1999).

39. Cf. chapters 2 and 3, as well as Hans-Jörg Rheinberger, "A History of Protein Biosynthesis and Ribosome Research," in *Protein Synthesis and Ribosome Structure: Translating the Genome*, ed. Knud H. Nierhaus and Daniel N. Wilson (Weinheim: Wiley-VCH, 2004), 1–51.

40. Cf., e.g., Nancy Nersessian, "Engineering Devices: The Culture of Physical Simulation Modeling in Biomedical Engineering," in *Cultures Without Culturalism: The Making of Scientific Knowledge*, ed. Karine Chemla and Evelyn Fox Keller (Durham, NC and London: Duke Univer-sity Press, 2017), 117–44.

41. See, e.g., Jean-Paul Gaudillière, *Inventer la biomédecine: La France, l'Amerique et la pro-duction des savoirs du vivant (1945–1965)* (Paris: Editions la Découverte, 2002); Peter Keating and Alberto Cambrosio, *Biomedical Platforms: Realigning the Normal and the Pathological in Late-Twentieth-Century Medicine* (Cambridge, MA: MIT Press, 2003).

42. See also Hans-Jörg Rheinberger, *Natur und Kultur im Spiegel des Wissens* (Heidelberg: Universitätsverlag Winter, 2015).

43. Karl Mannheim, *Structures of Thinking*, Vol. 10 of *The Collected Works of Karl Mannheim*, trans. Jeremy J. Shapiro and Sierry Webar Nicholsen (London: Routledge, 1991), 45–46.

44. Michel Serres, *The Natural Contract*, trans. Elizabeth MacArthur and William Paulson (Ann Arbor: University of Michigan Press, 1995), 3.

45. Serres, 71–72.

46. Andrew Pickering, ed., *Science as Practice and Culture* (Chicago: University of Chicago Press, 1992).

47. Knorr Cetina, *Epistemic Cultures*.

48. Karine Chemla and Evelyn Fox Keller, eds., *Cultures Without Culturalism: The Making of Scientific Knowledge* (Durham, NC and London: Duke University Press, 2017).

49. Mannheim, *Structures of Thinking*, 99.

50. Pierre Bourdieu, *Pascalian Meditations*, trans. Richard Nice (Stanford, CA: Stanford University Press, 2000), 109.

51. Gaston Bachelard, *Le Rationalisme appliqué* (Paris: Presses Universitaires de France, 1949), 132.

52. Gaston Bachelard, *The Philosophy of No: A Philosophy of the New Scientific Mind*, trans. G. C. Waterston (New York: Orion Press, 1968), 12, emphasis in the original, translation slightly altered.

53. Bachelard, *Rationalisme appliqué*, 4.

54. Bachelard, 43.

55. Bachelard, 132–33.

56. Bachelard, 43.

57. Bachelard, 133.

58. Bachelard, 133.

59. Bachelard, 31.

60. See, e.g., John Dupré, *The Disorder of Things: Metaphysical Foundations of the Dis-unity of Science* (Cambridge, MA: Harvard University Press, 1993); Peter Galison and David J. Stump, eds., *The Disunity of Science: Boundaries, Contexts, and Power* (Stanford, CA: Stanford

University Press, 1996); Ida H. Stamhuis et al., eds., *The Changing Images of Unity and Disunity in the Philosophy of Science* (Dordrecht: Kluwer, 2002); Stephen H. Kellert, Helen E. Longino, and C. Kenneth Waters, eds., *Scientific Pluralism* (Minneapolis: University of Minnesota Press, 2006).

61. Susan L. Star and James R. Griesemer, "Institutional Ecology, 'Translations' and Boundary Objects: Amateurs and Professionals in Berkeley's Museum of Vertebrate Zoology, 1907–1939," *Social Studies of Science* 19 (1989): 387–420.

Chapter 8

1. On narration in science, see Mary S. Morgan and Norton Wise, eds., "Narrative in Science," *Studies in History and Philosophy of Science, Part A* 62, special issue (2017): 1–98.

2. Hans Blumenberg, *Die Lesbarkeit der Welt* (Frankfurt am Main: Suhrkamp, 1981).

3. For Galileo, see Alexandre Koyré, *From the Closed World to the Open Universe* (Baltimore, MD: Johns Hopkins University Press, 1957); see also Hans Blumenberg, "Das Fernrohr und die Ohnmacht der Wahrheit," introduction to Galileo Galilei, *Sidereus Nuncius—Nachricht von neuen Sternen* (Frankfurt am Main: Suhrkamp, 1980), 7–75. For the graphical method, see François Dagognet, *Etienne Jules Marais: La Passion de la trace* (Paris: Hazan, 1987); Robert Brain, *The Pulse of Modernism: Physiological Aesthetics in Fin de Siècle Europe* (Spokane: University of Washington Press, 2015). For the genetic code see Wolfgang Raible, *Sprachliche Texte—Genetische Texte: Sprachwissenschaft und molekulare Genetik* (Heidelberg: Universitätsverlag Winter, 1993); see also Werner Kogge, *Experimentelle Begriffsforschung: Philosophische Interventionen am Beispiel von Code, Information und Skript in der Molekularbiologie* (Weilerswist: Velbrück, 2017).

4. Classically, Karl Popper, *The Logic of Scientific Discovery* (New York: Basic Books, 1959); actually, and critically, Ronald L. Numbers, ed., *Newton's Apple and Other Myths about Science* (Cambridge, MA: Harvard University Press, 2015).

5. On research machines, see Henning Schmidgen, *Forschungsmaschinen: Experimente zwischen Wissenschaft und Kunst* (Berlin: Matthes & Seitz, 2017).

6. Edmund Husserl, "The Origin of Geometry," in *Edmund Husserl's Origin of Geometry: An Introduction*, ed. Jacques Derrida (Lincoln and London: University of Nebraska Press, 1989); see also Friedrich Steinle, "Concepts, Facts, and Sedimentation in Experimental Science," in *Science and the Life-World: Essays on Husserl's Crisis of European Sciences*, ed. David Hyder and Hans-Jörg Rheinberger (Stanford, CA: Stanford University Press, 2010), 199–214.

7. Gaston Bachelard, *Le Rationalisme appliqué* (Paris: Presses Universitaires de France, 1949), 103.

8. Johannsen spoke of "geno-differences" as the epistemic objects of the genetic experimental systems of his time. See Wilhelm Johannsen, "The Genotype Conception of Heredity," *American Naturalist* 45 (1911):129–59, esp. 150.

9. Bachelard, *Rationalisme appliqué*, 4.

10. Husserl, "Origin of Geometry," 172.

11. Husserl, 174.

12. László Tengelyi, *Welt und Unendlichkeit: Zum Problem phänomenologischer Metaphysik* (Freiburg and Munich: Karl Alber, 2014), 218.

13. Jacques Derrida, *Edmund Husserl's Origin of Geometry: An Introduction*, trans. John P. Leavey (Lincoln and London: University of Nebraska Press, 1989), 150.

14. Husserl, "Origin of Geometry", 158–59.

15. Cf. Gilles Deleuze, *Difference and Repetition*, trans. Paul Patton (New York: Columbia University Press, 1994).

16. Cf. Paul Zamecnik, "Historical and Current Aspects of the Problem of Protein Synthesis," in *The Harvey Lectures 1958–59* (New York and London: Academic Press, 1960), 256–81.

17. Paul Zamecnik, "Historical Aspects of Protein Synthesis," *Annals of the New York Academy of Sciences* 325 (1979): 269–301.

18. Gaston Bachelard, *L'Activité rationaliste de la physique contemporaine* (Paris: Presses Universitaires de France, 1951), 51.

19. Bachelard, *Rationalisme appliqué*, 2, emphasis in original.

20. Bachelard, *Activité rationaliste*, 27.

21. Bachelard, 25, emphasis in original.

22. Hayden White, *The Content of the Form: Narrative Discourse and Historical Representation* (Baltimore, MD and London: Johns Hopkins University Press, 1987), 2.

23. Emile Benveniste, *Problems in General Linguistics*, trans. Mary Elizabeth Meek (Coral Gables, FL: University of Miami Press, 1973), 208.

24. White, *Content of the Form*, 51; cf. Paul Ricœur, "Narrative Time," *Critical Inquiry* 7 (1980): 169–90, as well as Paul Ricœur, *Time and Narrative*, Vol. 1, trans. Kathleen McLaughlin and David Pellauer (Chicago: University of Chicago Press, 1984).

25. Tristan Tzara, *Grains et issues* (Paris: Garnier-Flammarion, 1981), 155–56.

26. Tzara, 155.

27. Gaston Bachelard, "Le Surrationalisme," *Inquisitions: Organe du groupe d'études pour la phénoménologie humaine* 1 (1936): 1–6.

28. Gaston Bachelard, *The Poetics of Reverie: Childhood, Language, and the Cosmos*, trans. Daniel Russell (Boston: Beacon Press, 1969); see also Hans-Jörg Rheinberger, *The Hand of the Engraver: Albert Flocon Meets Gaston Bachelard*, trans. Kate Sturge (Albany: State University of New York Press, 2018).

29. Michael Polanyi, *Duke Lectures* (1964), microfilm, Library Photographic Service, University of California, Berkeley 1965, 4th lecture, 4–5, in Marjorie Grene, *The Knower and the Known* (Washington, DC: Center for Advanced Research in Phenomenology and University Press of America, 1984), 219.

30. Michael Polanyi, "The Unaccountable Element in Science," in *Knowing and Being: Essays by Michael Polanyi*, ed. Marjorie Grene (Chicago: University of Chicago Press, 1969), 119–20.

31. Gaston Bachelard, *The New Scientific Spirit*, trans. Arthur Goldhammer (Boston: Beacon Press, 1984).

32. Brian Rotman, *Signifying Nothing: The Semiotics of Zero* (New York: St. Martin's Press, 1987).

33. Ian Hacking, *Representing and Intervening: Introductory Topics in the Philosophy of Natural Science* (Cambridge: Cambridge University Press, 1983), in particular the chapter "Break," 130–46.

34. Bachelard, *Activité rationaliste*, 25.

35. Polanyi, "Unaccountable Element," 119.

36. Hans Blumenberg, *Theorie der Unbegrifflichkeit* (Frankfurt am Main: Suhrkamp, 2007).

37. Jean Cavaillès, *On Logic and the Theory of Science*, trans. Knox Peden and Robin Mackay (Cambridge, MA: MIT Press, 2021), 135–36.

38. Michael Polanyi, "The logic of tacit inference" (1964), in *Knowing and Being: Essays by Michael Polanyi*, ed. Marjorie Grene (Chicago: University of Chicago Press, 1969), 138–58.

39. Claude Bernard, *Cahier de notes 1850–1860*, ed. Mirko Grmek (Paris: Gallimard, 1965), 196.

40. Bernard, 152.

41. Bernard, 124.

42. Bernard, 135.

43. Robert K. Merton and Elinor Barber, *The Travels and Adventures of Serendipity: A Study in Sociological Semantic and the Sociology of Science* (Princeton, NJ: Princeton University Press, 2006). As we can see, the book's title subsumes its own subject matter under the principle of serendipity.

44. Robert K. Merton, *Social Theory and Social Structure* (New York: The Free Press, 1968), 157–62.

45. Popper, *Logic of Scientific Discovery*; Hans Reichenbach, *Experience and Prediction: An Analysis of the Foundations of the Structure of Knowledge* (Chicago: University of Chicago Press, 1938). See also, more recently, Jutta Schickore and Friedrich Steinle, eds., *Revisiting Discovery and Justification: Historical and Philosophical Perspectives on the Context Distinction* (Dordrecht: Springer, 2006); Allan Franklin and Ephraim Fischbach, *The Rise and Fall of the Fifth Force: Discovery, Pursuit, and Justification in Modern Physics*, 2nd edition (Cham: Springer, 2016).

46. Note the subtitle of Bruno Latour, *Science in Action: How to Follow Scientists and Engineers Through Society* (Cambridge, MA: Harvard University Press, 1987).

47. Cf. Hans-Jörg Rheinberger, *Toward a History of Epistemic Things: Synthesizing Proteins in the Test Tube* (Stanford, CA: Stanford University Press, 1997); Hans-Jörg Rheinberger, *An Epistemology of the Concrete: Twentieth-Century Histories of Life* (Durham, NC and London: Duke University Press, 2010b).

48. Georges Canguilhem, "The Object of the History of Sciences," in *Continental Philosophy of Science*, ed. Gary Gutting (Oxford: Blackwell Publishing, 2005), 203.

49. It is remarkable in this context that there are attempts to recruit the concept of "biography" as a metaphor for the microhistory of objects of knowledge. See Lorraine Daston, ed., *Biographies of Scientific Objects* (Chicago: University of Chicago Press, 2000).

50. Carlo Ginzburg, "Microhistory," in *Threads and Traces: True False Fictive*, trans. Anne C. Tedeschi and John Tedeschi (Berkeley, Los Angeles, and London: University of California Press, 2012), 209–10.

51. For this concept, see Andrew Pickering, *The Mangle of Practice: Time, Agency, and Science* (Chicago: University of Chicago Press, 1995).

52. Cf. Andreas Cremonini and Markus Klammer, eds., *Bild-Beispiele: Zu einer pikturalen Logik des Exemplarischen* (Munich: Fink, 2020).

53. Giovanni Levi, "On Microhistory," in *New Perspectives on Historical Writing*, 2nd edition, ed. Peter Burke (University Park: Pennsylvania State University Press, 2001), 101–2.

54. Ginzburg, "Microhistory," 212–13.

55. George Kubler, *The Shape of Time. Remarks on the History of Things* (New Haven, CT and London: Yale University Press, 1962), 101–2.

56. The literature abounds with examples. As a landmark, compare Arthur O. Lovejoy, *The Great Chain of Being: A Study of the History of an Idea* (Cambridge, MA: Harvard University Press, 1936); for the life sciences as a whole, see Ernst Mayr, *The Growth of Biological Thought: Diversity, Evolution, and Inheritance* (Cambridge, MA: The Belknap Press of Harvard University Press, 1982).

57. Histories of disciplines are ubiquitous. For a metahistory, see Rudolf Stichweh, *Zur Entstehung des modernen Systems wissenschaftlicher Disziplinen—Physik in Deutschland 1740–1890*

(Frankfurt am Main: Suhrkamp, 1984); see also Timothy Lenoir, ed., *Instituting Science: The Cultural Production of Scientific Disciplines* (Stanford, CA: Stanford University Press, 1997); on the history of science itself as a discipline, see, e.g., Tore Frängsmyr, *History of Science in Sweden: The Growth of a Discipline, 1932–1982* (Uppsala: University of Uppsala, 1985).

58. Compare the classic study of François Jacob, *The Logic of Life: A History of Heredity*, trans. Betty E. Spillmann (New York: Pantheon Books, 1973); see also Christina Brandt, *Metapher und Experiment: Von der Virusforschung zum genetischen Code* (Göttingen: Wallstein, 2004); as well as Rheinberger, *Epistemology of the Concrete*, chapter 10.

59. Cf. Lily E. Kay, *Who Wrote the Book of Life? A History of the Genetic Code* (Stanford, CA: Stanford University Press, 2000).

60. Cf. Soraya de Chadarevian and Hans-Jörg Rheinberger, eds., "Disciplinary Histories and the History of Disciplines: The Challenge of Molecular Biology," *Studies in History and Philosophy of Biological and Biomedical Sciences* 40, no. 1, special issue (2009).

61. Stephen Hilgartner, *Reordering Life: Knowledge and Control in the Genomics Revolution* (Cambridge, MA: MIT Press, 2017).

62. See, generally, Eric J. Vettel, *Biotech: The Countercultural Origins of an Industry* (Philadelphia: University of Pennsylvania Press, 2006); for case studies see Nicolas Rasmussen, *Gene Jockeys: Life Science and the Rise of Biotech Enterprise* (Baltimore, MD: Johns Hopkins University Press, 2014); Doogab Yi, *The Recombinant University: Genetic Engineering and the Emergence of Stanford Biotechnology* (Chicago: University of Chicago Press, 2015).

63. For more detail, see Hans-Jörg Rheinberger, *Natur und Kultur im Spiegel des Wissens* (Heidelberg: Universitätsverlag Winter, 2015).

64. Bachelard, *Rationalisme appliqué*, 132–33.

65. See, as a summary, Pierre Bourdieu, *Pascalian Meditations*, trans. Richard Nice (Stanford, CA: Stanford University Press, 2000).

66. Michel Foucault, *The Order of Things: An Archaeology of the Human Sciences* (New York: Pantheon Books, 1970); Michel Foucault, *The Archaeology of Knowledge & The Discourse on Language*, trans. A. M. Sheridan Smith and Rupert Swyer (New York: Pantheon Books, 1972).

67. Foucault, *Discourse on Language*, 235.

68. Georges Canguilhem, *La Formation du concept de réflexe aux XVIIᵉ et XVIIIᵉ siècles* (Paris: Presses Universitaires de France, 1955); see also Henning Schmidgen, "The Life of Concepts: Georges Canguilhem and the History of Science," *History and Philosophy of the Life Sciences* 36 (2014c): 232–53.

69. Staffan Müller-Wille and Hans-Jörg Rheinberger, *A Cultural History of Heredity* (Chicago: University of Chicago Press, 2012).

70. Ohad Parnes, Ulrike Vedder, and Stefan Willer, *Generation: Eine Geschichte der Wissenschaft und der Kultur* (Frankfurt am Main: Suhrkamp, 2008).

71. For the concept of "image of science," see Yehuda Elkana, "Programmatic Attempt at an Anthropology of Knowledge," in *Sciences and Cultures: Anthropological and Historical Studies*, ed. Everett Mendelsohn and Yehuda Elkana (Dordrecht: Reidel, 1981), 1–76.

72. Jean-François Lyotard, *The Postmodern Condition: A Report on Knowledge*, trans. Geoffrey Bennington and Brian Massumi (Minneapolis: University of Minnesota Press, 1984).

73. Cf. Georges Canguilhem, *Ideology and Rationality in the History of Life Sciences*, trans. Arthur Goldhammer (Cambridge, MA: MIT Press, 1988); on Canguilhem's usage of the concept of "scientific ideology," see Claude Debru, *Georges Canguilhem, science et non-science* (Paris: Editions Rue d'Ulm/Presses de l'Ecole normale supérieure, 2004), in particular chapter 4.

Chapter 9

1. François Jacob, *Of Flies, Mice, and Men*, trans. Giselle Weiss (Cambridge, MA and London: Harvard University Press, 1998), 126.

2. Hence the French title of the book (*Le Jeu des possibles*). See François Jacob, *The Possible and the Actual*, trans. Betty E. Spillmann (Seattle: University of Washington Press, 1982).

3. François Jacob, *The Logic of Life: A History of Heredity*, trans. Betty E. Spillmann (New York: Pantheon Books, 1973), 324.

4. Claude Lévi-Strauss, *Introduction to a Science of Mythology I: The Raw and the Cooked*, trans. John and Doreen Weightman (New York: Harper and Row, 1969), 7.

5. Lévi-Strauss, 2–3.

6. Claude Lévi-Strauss, *Wild Thought*, trans. Jeffrey Mehlman and John Leavitt (Chicago: University of Chicago Press, 2021), 16.

7. Lévi-Strauss, *Wild Thought*, 3, 18.

8. Lévi-Strauss, 20.

9. Lévi-Strauss, *The Raw and the Cooked*, 11n3. There, Lévi-Strauss quotes Paul Ricoeur's "Symbole et temporalité" (*Archivio di Filosofía* 1–2 [1963]: 9–10), and refers to Roger Bastide's "La Nature humaine: Le point de vue du sociologue et de l'ethnologue," in *La Nature humaine: Actes du XIe Congrès des sociétés de philosophie de langue française (Montpellier, 4–6 septembre 1961)* (Paris: Presses Universitaires de France, 1961), 65–79.

10. Lévi-Strauss, *Wild Thought*, title of chapter I.

11. Ernst Cassirer, *Philosophy of Symbolic Forms*, Vol. 2, *Mythical Thought*, trans. Ralph Manheim (New Haven, CT: Yale University Press, 1953–1957).

12. I have dealt more comprehensively with Bachelard's reflections on a dialectic of abstraction and concretion in my book *The Hand of the Engraver*, trans. Kate Sturge (Albany: State University of New York Press, 2018), devoted to Bachelard's collaboration with the copper engraver Albert Flocon,

13. Gaston Bachelard, *Le Rationalisme appliqué* (Paris: Presses Universitaires de France, 1949), 1, emphasis in original. See also Jean Gayon, *Gaston Bachelard: le raionalisme appliqué* (Paris: Centre National d'Enseignement à Distance/Presses Universitaires de France, 1994).

14. Bachelard, *Rationalisme appliqué*, 3.

15. Bachelard, 50–51.

16. Bachelard, 10, emphasis in original.

17. Bachelard, 43.

18. Gaston Bachelard, "Le Surrationalisme," *Inquisitions: Organe du groupe d'études pour la phénoménologie humaine* 1 (1936): 1–6.

19. Edgar Wind, *Experiment and Metaphysics: Towards a Resolution of the Cosmological Antinomies*, trans. Cyril Edwards (Oxford: European Humanities Research Centre of the University of Oxford, 2001), 9; for Wind, see also Horst Bredekamp et al., eds., *Edgar Wind: Kunsthistoriker und Philosoph* (Berlin: Akademie Verlag, 1999).

20. Wind, *Experiment and Metaphysics*, 19.

21. Bachelard, *Rationalisme appliqué*, 9. Jean-Toussaint Desanti has resumed this thread of a critique of the "science of the philosophers." Cf. Jean-Toussaint Desanti, *La Philosophie silencieuse, ou critique des philosophies de la science* (Paris: Editions du Seuil, 1975).

22. Bachelard, *Rationalisme appliqué*, 10.

23. Bachelard, 43.

24. Bachelard, 11, emphasis in original.

25. Bachelard, 54, emphasis in original.

26. Bachelard, 51.

27. Lévi-Strauss, *Wild Thought*, 284.

28. Ernst Cassirer, *The Problem of Knowledge: Philosophy, Science and History since Hegel*, trans. William H. Woglom and Charles W. Hendel (New Haven, CT: Yale University Press, 1950), 18.

29. Cf. Gaston Bachelard, *The Philosophy of No: A Philosophy of the New Scientific Mind*, trans. G. C. Waterston (New York: Orion Press, 1968), in particular the preface.

30. Gaston Bachelard, *The New Scientific Spirit*, trans. Arthur Goldhammer (Boston: Beacon Press, 1984), 172 (among others).

31. Lévi-Strauss, *Wild Thought*, 20–21.

32. As an example, I refer to the remarks on the history of electron microscopy in chapter 5 above, and the literature quoted there.

33. See, e.g., Maureen A. O'Malley, "Making Knowledge in Synthetic Biology: Design Meets Kludge," *Biological Theory* 4 (2009): 378–89; Karen Kastenhofer, "Synthetic Biology as Understanding, Control, Construction, and Creation? Techno-epistemic and Socio-political Implications of Different Stances in Talking and Doing Technoscience," *Futures* 48 (2013): 13–22.

34. Wind, *Experiment and Metaphysics*, 21–22, emphasis in original.

35. Gaston Bachelard, *The Formation of the Scientific Mind: A Contribution to a Psychoanalysis of Objective Knowledge*, trans. Mary McAllester Jones (Manchester: Clinamen, 2002), chapter 1.

36. Bachelard, *Formation of the Scientific Mind*, 24.

37. Bachelard, 24.

38. Paul Feyerabend, *Against Method*, 3rd edition (London and New York: Verso, 1993), 17.

39. Paul Feyerabend, *Wider den Methodenzwang: Skizze einer anarchistischen Erkenntnistheorie* (Frankfurt am Main: Suhrkamp, 1976), 39, emphasis in original.

40. Bachelard, *Formation of the Scientific Mind*, chapter 12.

41. Hélène Metzger, *La Méthode philosophique en histoire des sciences: Textes 1914–1939* (Paris: Fayard, 1987); see also Gad Freudenthal, ed., *Etudes sur/Studies on Hélène Metzger* (Leiden: Brill, 1990).

42. Cf. Samuel Talcott, *Georges Canguilhem and the Problem of Error* (London: Palgrave Macmillan, 2019).

43. Federigo Enriques, *Signification de l'histoire de la pensée scientifique* (Paris: Editions Hermann, 1934), 17.

44. Gaston Bachelard, *The Poetics of Reverie: Childhood, Language, and the Cosmos*, trans. Daniel Russell (Boston: Beacon Press, 1969), 3. See also Timo Kehren et al., eds., *Staunen: Perspektiven eines Phänomens zwischen Natur und Kultur* (Paderborn: Fink, 2019), especially the contribution of Toni Hildebrandt, "Postapokalyptisches Staunen: Ästhetik und Geschichtsbewusstsein im Naturvertrag," 237–59.

45. André Darbon, *Une Philosophie de l'expérience* (Paris: Presses Universitaires de France, 1946), 113.

46. Bachelard, *Formation of the Scientific Mind*, 22.

47. Hence the subtitle of *Formation of the Scientific Mind*.

48. Bachelard, *Formation of the Scientific Mind*, 86.

49. Bachelard, 81.

50. Bachelard, 82.

51. Lévi-Strauss, *Wild Thought*, 23.

52. Lévi-Strauss, 26.

53. Lévi-Strauss, 26.

54. Lévi-Strauss, 282.

55. Bruno Latour, *We Have Never Been Modern*, trans. Catherine Porter (Cambridge, MA: Harvard University Press, 1993), 142.

Chapter 10

1. See, e.g., Reinhard Brunner, *Die Fragmentierung der Vernunft: Rationalitätskritik im 20. Jahrhundert* (Frankfurt am Main: Campus, 1994); François Audigier, Anne Sgard, and Nicole Tutiaux-Guillon, eds., *Sciences de la nature et de la société dans une école en mutation: Fragmentations, recompositions, nouvelles alliances?* (Louvain-la-Neuve: De Boeck Supérieur, 2015); for a critical approach to the fragment, see also Michel Serres, *Conversations on Science, Culture, and Time: Michel Serres Interviewed by Bruno Latour*, trans. Roxanne Lapidus (Ann Arbor: University of Michigan Press, 1995), 119–22.

2. See chapter 8 above. See also Edmund Husserl, "The Origin of Geometry," in *Edmund Husserl's Origin of Geometry: An Introduction*, ed. Jacques Derrida (Lincoln and London: University of Nebraska Press, 1989), 158–59.

3. François Jacob, *The Statue Within: An Autobiography*, trans. Franklin Philip (New York: Basic Books, 1988), 234.

4. Peter Galison, *Image & Logic: A Material Culture of Microphysics* (Chicago: University of Chicago Press, 1997).

5. Concerning its genesis, see Wolf von Engelhardt, "'Der Versuch als Vermittler von Objekt und Subjekt': Goethes Aufsatz im Licht von Kants Vernunftkritik," *Athenäum: Jahrbuch für Romantik* 10 (2000): 9–28.

6. First published in Johann Wolfgang von Goethe, *Zur Naturwissenschaft überhaupt*, 2. Band, 1. Heft (Tübingen and Stuttgart: Cotta, 1823).

7. Johann Wolfgang von Goethe, "The Experiment as Mediator of Object and Subject," trans. Craig Holdrege, *Context* 24 (Fall 2010): 20.

8. Goethe, "Experiment as Mediator," 22.

9. For good examples, see Elliott Sober, "Independent Evidence About a Common Cause," *Philosophy of Science* 56 (1989): 275–87; Jacob Stegenga and Tarun Menon, "Robustness and Independent Evidence," *Philosophy of Science* 84 (2017): 424–35.

10. Stanislav V. Borisov and Nina V. Podberezskaya, "X-ray Diffraction Analysis: A Brief History and Achievements of the First Century," *Journal of Structural Chemistry* 53 (2012): 1–3.

11. Fred L. Suddath et al., "Three-dimensional Structure of Phenylalanine Transfer RNA at 3.0 Å Resolution," *Nature* 248 (1974): 20–24.

12. Compare Carsten Reinhardt, *Shifting and Rearranging: Physical Methods and the Transformation of Modern Chemistry* (Sagamore Beach, MA: Science History Publications/USA, 2006).

13. For a contemporary overview, see Pablo Fernández-Millán et al., "Transfer RNA: From Pioneering Crystallographic Studies to Contemporary tRNA Biology," *Archives of Biochemistry and Biophysics* 602 (2016): 95–105.

14. Cf. Viter Marquéz and Knud H. Nierhaus, "tRNA and Synthetases," in *Protein Synthesis and Ribosome Structure: Translating the Genome*, ed. Knud H. Nierhaus and Daniel N. Wilson (Weinheim: Wiley-VCH, 2004), esp. 149–54.

15. Georges Canguilhem, "The Role of Analogies and Models in Biological Discovery," in *Scientific Change*, ed. Alistair C. Crombie (London: Heinemann, 1963), 513.

16. Claude Lévi-Strauss, *Structural Anthropology*, trans. Claire Jacobson and Grooke Grundfest Schoepf (New York: Basic Books, 1963), 279.

17. The locus classicus is Max Black, *Models and Metaphors* (Ithaca, NY: Cornell University Press, 1962). For molecular biology, cf. also Christina Brandt, *Metapher und Experiment: Von der Virusforschung zum genetischen Code* (Göttingen: Wallstein, 2004).

18. Compare, in an exemplary fashion, Carlo Ginzburg, *Clues, Myths and the Historical Method*, trans. John Tedeschi and Anne C. Tedeschi (Baltimore, MD: Johns Hopkins University Press, 1989); Carlo Ginzburg, *Threads and Traces: True False Fictive*, trans. Anne C. Tedeschi and John Tedeschi (Berkeley, Los Angeles, and London: University of California Press, 2012).

19. Cf. Martin Rudwick, *The Meaning of Fossils: Episodes in the History of Palaeontology* (London: MacDonald, 1972); Stephen J. Gould, *Time's Arrow, Time's Cycle* (Cambridge, MA: Harvard University Press, 1987). For a case study from paleoanthropology, see Marianne Sommer, *Bones and Ochre: The Curious Afterlife of the Red Lady of Paviland* (Cambridge, MA: Harvard University Press, 2007).

20. Cf. David Sepkoski, *Rereading the Fossil Record: The Growth of Paleobiology as an Evolutionary Discipline* (Chicago: University of Chicago Press, 2012).

21. Martin Rudwick, *Scenes from Deep Time: Early Pictorial Images of the Prehistoric World* (Chicago: University of Chicago Press, 1992).

22. Charles Lyell, *A Manual of Elementary Geology; or, The Ancient Changes on the Earth and Its Inhabitants as Illustrated by Geological Monuments* (London: John Murray, 1851).

23. Georges Cuvier, *Essay on the Theory of the Earth*, trans. Robert Kerr (New York: Kirk & Mercein, 1818), 102–3; see also Martin Rudwick, *Georges Cuvier, Fossil Bones, and Geological Catastrophes* (Chicago: University of Chicago Press, 1997).

24. See, e.g., Ulrich Veit, "Archäologiegeschichte als Wissenschaftsgeschichte: Über Formen und Funktionen historischer Selbstvergewisserung in der Prähistorischen Archäologie," *Ethnographisch-Archäologische Zeitschrift* 52 (2011): 34–58; Sabine Reinhold and Kerstin P. Hofmann, eds., "Zeichen der Zeit: Archäologische Perspektiven auf Zeiterfahrung, Zeitpraktiken und Zeitkonzepte," *Forum Kritische Archäologie* 3, thematic issue (2014); Undine Stabrey, *Archäologische Untersuchungen: Über Temporalität und Dinge* (Bielefeld: Transcript, 2017).

25. Ludwik Fleck, "Zur Krise der Wirklichkeit," in *Erfahrung und Tatsache: Gesammelte Aufsätze* (Frankfurt am Main: Suhrkamp, 1983), 46–58.

26. Fleck, "Krise der Wirklichkeit," 50.

27. Michel Serres, "Introduction," in *A History of Scientific Thought: Elements of a History of Science*, ed. Michel Serres (Oxford: Blackwell, 1995); see also Serres, *Conversations*, 57–61.

28. Compare, e.g., Otto Sibum, "Reworking the Mechanical Value of Heat: Instruments of Precision and Gestures of Accuracy in Early Victorian England," *Studies in History and Philosophy of Science Part A* 26 (1995): 73–106; Peter Heering, Falk Rieß, and Christian Sichau, eds., *Im Labor der Physikgeschichte: Zur Untersuchung historischer Experimentalpraxis* (Oldenburg: BIS-Verlag, 2000).

29. Compare the journal *Annales d'histoire économique et sociale*, founded in 1929 by Marc Bloch and Lucien Febvre, since 1994 *Annales: Histoire, Sciences sociales*; most recently Romain Bertrand and Guillaume Calafat, eds., "Micro-analyse et histoire globale," *Annales* 73, no. 1, special issue (2019).

30. Paradigmatically, Carlo Ginzburg, *The Cheese and the Worms: The Cosmos of a Sixteenth-Century Miller*, trans. John Tedeschi and Anne C. Tedeschi (New York: Penguin Books, 1982);

Giovanni Levi, *L'Eredità immateriale: Carriera di un esorcista ne Piemonte del Seicento* (Turin: Einaudi, 1985); David Warren Sabean, *Property, Production and Famliy in Neckarhausen, 1700–1870* (Cambridge: Cambridge University Press, 1990); Hans Medick, *Weben und Überleben in Laichingen, 1650–1900: Lokalgeschichte als allgemeine Geschichte* (Göttingen: Vandenhoeck & Ruprecht, 1996); see also Winfried Schulze, ed., *Sozialgeschichte, Alltagsgeschichte, Mikro-Historie: Eine Diskussion* (Göttingen: Vandenhoeck & Ruprecht, 1994).

Postscript

1. Lúcio Alberto Pinheiro dos Santos, *La Rythmanalyse* (Rio de Janeiro: Société de Psychologie et de Philosophie, 1931).

2. Gaston Bachelard, *The Dialectic of Time*, trans. Mary McAllester Jones (Manchester: Clinamen Press, 2000), 138.

3. Gaston Bachelard, "Castles in Spain," in *The Right to Dream*, trans. J. A. Underwood (New York: Orion, 1971), 83.

4. Henri Lefèbvre, *Rhythmanalysis: Space, Time and Everyday Life*, trans. Stuart Elden and Gerald Moore (London and New York: Continuum, 2004), 6–7.

5. Bachelard, *Dialectic of Time*, 20.

Bibliography

For every chapter, there exist a number of smaller preliminary papers to which the text does not refer in detail. They are listed below.

Chapter 1: "Auto-radio-graphics," in *Iconoclash*, ed. Bruno Latour and Peter Weibel (Cambridge, MA: MIT Press, 2002), 516–19; "Wie werden aus Spuren Daten, und wie verhalten sich Daten zu Fakten?" *Nach Feierabend: Zürcher Jahrbuch für Wissensgeschichte* 3 (Zurich and Berlin: Diaphanes, 2007), 117–25; "Spurenlesen im Experimentalsystem," in *Spur: Spurenlesen als Orientierungstechnik und Wissenskunst*, ed. Sybille Krämer, Werner Kogge, and Gernot Grube (Frankfurt am Main: Suhrkamp, 2007), 293–308; "Infra-experimentality: From Traces to Data, from Data to Patterning Facts," *History of Science* 49 (2011): 337–48; "Transpositions: From Traces through Data to Models and Simulations," in *Transpositions: Aesthetico-Epistemic Operators in Artistic Research*, ed. Michael Schwab (Leuven: Leuven University Press, 2018), 215–24.

Chapter 2: "A History of Protein Biosynthesis and Ribosome Research," in *Protein Synthesis and Ribosome Structure: Translating the Genome*, ed. Knud H. Nierhaus and Daniel N. Wilson (Weinheim: Wiley-VCH, 2004), 1–51; "Molekulare Modelle als epistemische Objekte. Ribosomen im Spiegel von 50 Jahren Forschung," in *Weltwissen. 300 Jahre Wissenschaften in Berlin*, ed. Jochen Hennig and Udo Andraschke (Munich: Hirmer Verlag, 2010), 75–81; "Preparations, Models, and Simulations," *History and Philosophy of the Life Sciences* 36 (2015): 321–34; "Über den Eigensinn epistemischer Dinge," in *Vom Eigensinn der Dinge. Für eine neue Perspektive auf die Welt des Materiellen*, ed. Hans Peter Hahn (Berlin: Neofelis Verlag, 2015), 147–62.

Chapter 3: "Epistemologica: Präparate," in *Das Museum als Erkenntnisort*, ed. Anke te Heesen and Petra Lutz (Cologne, Weimar, and Vienna: Böhlau, 2005), 65–75; "Sichtbar Machen. Visualisierung in den Naturwissenschaften," in *Bildtheorien. Anthropologische und kulturelle Grundlagen des Visualistic Turn*, ed. Klaus Sachs-Hombach (Frankfurt am Main: Suhrkamp, 2009), 127–45; "Making Visible: Visualization in the Sciences—and in Exhibitions?" in *Preprint No. 399* (Berlin: Max Planck Institute for the History of Science, 2010), 9–23; "Einige Bemerkungen zu Moden in der Wissenschaft," in *Mode und Moden*, ed. Heini Bader, Olaf Knellessen, Tamara Lewin, Angelika Oberhauser, and Husam Suliman (Zurich: Seismo Verlag, 2017), 79–91.

Chapter 4: "Pfropfen in Experimentalsystemen," in *Impfen, Pfropfen, Transplantieren*, ed. Uwe Wirth and Veronika Sellier (Berlin: Kadmos, 2011), 65–74; "On the Language of the History of Science," in *Handbook Language—Culture—Communication*, HSK 43, ed. Ludwig Jäger, Werner Holly, Peter Krapp, Samuel Weber, and Simone Heekeren (Berlin: De Gruyter, 2016), 341–47; "Schreiben und Experimentieren," in *Die Dringlichkeit der Literatur, Essays, Jahrbuch* 12 (Triesen: Literaturhaus Liechtenstein, 2018), 117–24.

Chapter 5: "Acht Miszellen zur Notation in den Wissenschaften," in *Notation. Kalkül und Form in den Künsten*, ed. Hubertus von Amelunxen, Dieter Appelt, and Peter Weibel (Berlin/Karlsruhe: Akademie der Künste/ZKM, 2008), 279–88; "Zettelwirtschaft," in *Schreiben als Kulturtechnik*, ed. Sandro Zanetti (Frankfurt am Main: Suhrkamp, 2012), 441–52.

Chapter 6: "Die Mehlmotte," in *Zoologicon. Ein kulturhistorisches Wörterbuch der Tiere*, ed. Christian Kassung, Jasmin Mersmann, and Olaf B. Rader (Munich: Fink, 2012), 259–62; "George Kubler und die Wissenschaftsgeschichte," in *Im Maschenwerk der Kunstgeschichte. Eine Revision von George Kublers "The Shape of Time,"* ed. Sarah Maupeu, Kerstin Schankweiler, and Stefanie Stallschus (Berlin: Kadmos, 2014), 245–50; "Experimentelle Serialität in Wissenschaft und Kunst," in *Serialität. Wissenschaften, Künste, Medien*, ed. Olaf Knellessen, Giaco Schiesser, and Daniel Strassberg (Berlin: Neofelis Verlag, 2015), 68–77; "George Kubler and the Question of Time and Temporality," in *Vision in Motion: Streams of Sensation and Configurations of Time*, ed. Michael Zimmermann (Zurich and Berlin: Diaphanes, 2016), 549–56.

Chapter 7: "Kulturen des Experiments," *Berichte zur Wissenschaftsgeschichte* 30 (2007): 135–44; "In vitro," in *UnTot. Existenzen zwischen Leben und Leblosigkeit*, ed. Peter Geimer (Berlin: Kadmos, 2014), 68–79; *Natur und Kultur im Spiegel des Wissens* (Heidelberg: Universitätsverlag, Winter 2015); "Cultures of Experimentation," in *Cultures without Culturalism: The Making of Scientific Knowledge*, ed. Karine Chemla and Evelyn Fox Keller (Durham, NC and London: Duke University Press, 2017), 278–95.

Chapter 8: "Noch etwas über die experimentelle Ordnung der Dinge," in *Wissenschaft und Welterzählung: Fakt & Fiktion. Die narrative Ordnung der Dinge*, ed. Matthias Michel (Zurich: Chronos Verlag, 2013), 270–71; "On Epistemic Objects, and Around," in *Witte de With Review: Arts, Culture, and Journalism in Revolt*, vol. 1, ed. Defne Aya and Adam Kleinman (Rotterdam: Witte de With Publishers, 2017), 376–81; "On the Narrative Order of Experimentation," in *Narratives and Comparisons: Adversaries or Allies in Understanding Science?* ed. Martin Carrier, Rebecca Mertens, and Carsten Reinhardt (Bielefeld: Bielefeld University Press, 2021), 85–97.

Chapter 9: "Das Wilde im Zentrum der Wissenschaft," interview with Wolfert von Rahden, *Gegenworte: Hefte für den Disput über Wissen* 12 (Berlin: Berlin-Brandenburgische Akademie der Wissenschaften, 2003), 36–38; "Die Wissenschaft des Konkreten," in *Iterationen* (Berlin: Merve, 2005), 101–28; "Das Wilde Denken," in *Zum Zufall*, Schriftenreihe, vol. 5 (Berlin: Willms Neuhaus Stiftung, 2018), 77–85; "A Remark on Gaston Bachelard's Idea of a Psychoanalysis of Knowledge," *Bachelard Studies* 2 (2021): 173–84; "Zum Beispiel Albert Flocon," in *Bild-Beispiele. Zu einer pikturalen Logik des Exemplarischen*, ed. Andreas Cremonini and Markus Klammer (Munich: Fink, 2020), 75–84.

Chapter 10: "Kleine Epistemologie des Fragments," in *Zum Zufall*, Schriftenreihe, vol. 6 (Berlin: Willms Neuhaus Stiftung, 2019), 93–103.

Abir-Am, Pnina. "From Biochemistry to Molecular Biology: DNA and the Acculturated Journey of the Critic of Science Erwin Chargaff." *History and Philosophy of the Life Sciences* 2 (1980): 3–60.

Abir-Am, Pnina. "The Politics of Macromolecules: Molecular Biologists, Biochemists, and Rhetoric." *Osiris* 7 (1992): 164–91.

Alberts, Bruce, Dennis Bray, Julian Lewis, Martin Raff, Keith Roberts, and James D. Watson. *Molecular Biology of the Cell*, 3rd edition. New York and London: Garland, 1994.

Alechinsky, Pierre. "Abstraction faite." *Cobra* 10 (Fall 1951): 4–8.

d'Alembert, Jean Le Rond. *Recherches sur differens points importans du système du monde*. Paris: Antoine Boudet & J. Chardon pour David l'aîné, 1754–1756.

Althusser, Louis. *Philosophy and the Spontaneous Philosophy of the Scientists and Other Essays*. Translated by Ben Brewster and others. London: Verso Books, 1989. Originally published as *Philosophie et philosophie spontanée des savants* (Paris: F. Maspero, 1974).

Ammon, Sabine. *Epistemologie des Entwerfens: Operative Artefakte und Visualisierungstechniken in Architektur und Ingenieurwesen*. Habilitationsschrift, Technische Universität Darmstadt, 2018.

André, Noël. *Théorie de la surface actuelle de la terre, ou plutôt Recherches impartiales sur le temps et l'agent de l'arrangement actuel de la surface de la terre, fondées, uniquement, sur les faits, sans système et sans hypothèse*. Paris: Société Typographique, 1806.

Ankeny, Rachel A. "Historiographic Reflections on Model Organisms; or How the Mureaucracy May Be Limiting Our Understanding of Contemporary Genetics and Genomics." *History and Philosophy of the Life Sciences* 32 (2010): 91–104.

Ankeny, Rachel A., and Sabina Leonelli. *Model Organisms*. Cambridge: Cambridge University Press, 2021.

Ankeny, Rachel A., and Sabina Leonelli. "What's So Special about Model Organisms?" *Studies in History and Philosophy of Science Part A* 42 (2011): 313–23.

Ascher, Jacques, and Jean-Pierre Jouet. *La Greffe, entre biologie et psychanalyse*. Paris: Presses Universitaires de France, 2004.

Audigier, François, Anne Sgard, and Nicole Tutiaux-Guillon, eds. *Sciences de la nature et de la société dans une école en mutation: Fragmentations, recompositions, nouvelles alliances?* Louvain-la-Neuve: De Boeck Supérieur, 2015.

Bachelard, Gaston. *L'Activité rationaliste de la physique contemporaine*. Paris: Presses Universitaires de France, 1951.

Bachelard, Gaston. "Castles in Spain." In *The Right to Dream*, trans. J. A. Underwood, 83–102. New York: Orion, 1971. Originally published as *Le Droit de rêver* (Paris: Presses Universitaires de France, 1970).

Bachelard, Gaston. *The Dialectic of Time*. Translated by Mary McAllester Jones. Manchester: Clinamen Press, 2000. Originally published as *La Dialectique de la durée* (Paris: Boivin & Cie., 1936).

Bachelard, Gaston. *The Formation of the Scientific Mind: A Contribution to a Psychoanalysis of Objective Knowledge*. Translated by Mary McAllester Jones. Manchester: Clinamen, 2002. Originally published as *La Formation de l'esprit scientifique: Contribution à une psychanalyse de la connaissance objective* (Paris: Vrin, 1938).

Bachelard, Gaston. *Le Matérialisme rationnel*. Paris: Presses Universitaires de France, 1953.

Bachelard, Gaston. *The New Scientific Spirit*. Translated by Arthur Goldhammer. Boston: Beacon Press, 1984. Originally published as *Le Nouvel esprit scientifique* (Paris: F. Alcan, 1934).

Bachelard, Gaston. "Noumenon and Microphysics." Translated by Bernard Roy. *Philosophical Forum* 37 (2006): 75–84. Originally published as "Noumène et microphysique," in *Recherches philosophiques*, vol. 1, ed. A. Koyré, H.-C. Puech, and A. Spaier, 55–65 (Paris: Boivin & Cie., 1931).

Bachelard, Gaston. *The Philosophy of No: A Philosophy of the New Scientific Mind*. Translated by G. C. Waterston. New York: Orion Press, 1968. Originally published as *La Philosophie du non* (Paris: Presses Universitaires de France, 1940).

Bachelard, Gaston. *The Poetics of Reverie: Childhood, Language, and the Cosmos*. Translated by Daniel Russell. Boston: Beacon Press, 1969. Originally published as *La Poétique de la rêverie* (Paris: Presses Universitaires de France, 1960).

Bachelard, Gaston. *Le Rationalisme appliqué*. Paris: Presses Universitaires de France, 1949.

Bachelard, Gaston. "Le Surrationalisme." *Inquisitions: Organe du groupe d'études pour la phénoménologie humaine* 1 (1936): 1–6.

Bachelard, Gaston. *Water and Dreams: An Essay on the Imagination of Matter*. Translated by Edith Farrell. Dallas: The Pegasus Foundation, 1983. Originally published as *L'Eau et les rêves* (Paris: J. Corti, 1942).

Badiou, Alain. *Le Concept de modèle*. Paris: Librairie Arthème Fayard, 2007. First published 1969 by F. Maspero (Paris).

Bal, Mieke. *Travelling Concepts in the Humanities: A Rough Guide*. Toronto: University of Toronto Press, 2002.

Ban, Nenad, Poul Nissen, Peter B. Moore, and Thomas A. Steitz. "Crystal Structure of the Large Ribosomal Subunit at 5-Ångstrom Resolution." In *The Ribosome: Structure, Function, Antibiotics, and Cellular Interactions*, ed. Roger Garrett, Stephen R. Douthwaite, Anders Liljas, Alistair T. Matheson, Peter B. Moore, and Harry E. Noller, 11–20. Washington: ASM Press, 2000.

Bashan, Anat, Marta Pioletti, Heike Bartels, Daniela Janell, Frank Schluenzen, Marco Gluehmann, Inna Levin, Joerg Harms, Harly A. S. Hansen, Ante Tocilji, Tamar Auerbach, Horacio Avila, Maria Simitsopoulou, Moshe Peretz, William S. Bennett, Ilana Agmon, Maggie

Kessler, Shulamith Weinstein, François Franceschi, and Ada Yonath. "Identification of Selected Ribosomal Compounds in Crystallographic Maps of Procaryotic Ribosomal Subunits at Medium Resolution." In *The Ribosome: Structure, Function, Antibiotics, and Cellular Interactions*, ed. Roger Garrett, Stephen R. Douthwaite, Anders Liljas, Alistair T. Matheson, Peter B. Moore, and Harry E. Noller, 21–33. Washington: ASM Press, 2000.

Bastide, Roger. "La Nature humaine: le point de vue du sociologue et de l'ethnologue." In *La Nature humaine: Actes du XIe Congrès des sociétés de philosophie de langue française (Montpellier, 4–6 septembre 1961)*, 65–79. Paris: Presses Universitaires de France, 1961.

Baudrillard, Jean. *Simulations*. Translated by Paul Foss, Paul Patton, and Philip Beitchman. New York: Semiotext(e), 1983.

Bauer, Matthias, and Christoph Ernst. *Diagrammatik: Einführung in ein kultur- und medienwissenschaftliches Forschungsfeld*. Bielefeld: transcript-Verlag, 2010.

Bechtel, William. *Discovering Complexity*. Princeton, NJ: Princeton University Press, 1993.

Bechtel, William. *Discovering Cell Mechanisms: The Creation of Modern Cell Biology*. Cambridge: Cambridge University Press, 2006.

Benveniste, Emile. *Problems in General Linguistics*. Translated by Mary Elizabeth Meek. Coral Gables, FL: University of Miami Press, 1973. Originally published as *Problèmes de linguistique générale* (Paris: Gallimard, 1966).

Berger, Silvia. *Bakterien in Krieg und Frieden: Eine Geschichte der medizinischen Bakteriologie in Deutschland 1890–1933*. Göttingen: Wallstein, 2009.

Bernard, Claude. *Cahier de notes 1850–1860*, complete edition of the "Cahier rouge." Edited by Mirko Dražen Grmek. Paris: Gallimard, 1965.

Bernard, Claude. *An Introduction to the Study of Experimental Medicine*. Translated by Henry Copley Greene. New York: Macmillan, 1927. Originally published as *Introduction à l'étude de la médecine expérimentale* (Paris: Baillière, 1865).

Bernard, Claude. *Lectures on the Phenomena of Life Common to Animals and Plants*. Translated by Hebbel E. Hoff, Roger Guillemin, and Lucienne Guillemin. Springfield, IL: Charles C. Thomas, 1974. Originally published as *Leçons sur les phénomènes de la vie communs aux animaux et aux végétaux* (Paris: Baillière, 1878/79).

Bernard, Claude. *Philosophie: Manuscrit inédit*. Edited by Jacques Chevalier. Paris: Editions Hatier-Boivin, 1954.

Bertrand, Romain, and Guillaume Calafat, eds. "Micro-analyse et histoire globale." *Annales* 73, no. 1, special issue (2019).

Black, Max. *Models and Metaphors*. Ithaca, NY: Cornell University Press, 1962.

Blumenberg, Hans. "Das Fernrohr und die Ohnmacht der Wahrheit." Introduction to *Sidereus Nuncius—Nachricht von neuen Sternen*, by Galileo Galilei, 7–75. Frankfurt am Main: Suhrkamp, 1980.

Blumenberg, Hans. *Die Lesbarkeit der Welt*. Frankfurt am Main: Suhrkamp, 1981.

Blumenberg, Hans. "Paradigmen zu einer Metaphorologie." *Archiv für Begriffsgeschichte* 6 (1960): 7–142, 301–5.

Blumenberg, Hans. *Phänomenologische Schriften 1981–1988*. Edited by Nicola Zambon. Berlin: Suhrkamp, 2018.

Blumenberg, Hans. *Theorie der Unbegrifflichkeit*. Edited by Anselm Haverkamp. Frankfurt am Main: Suhrkamp, 2007.

Bogen, James, and James Woodward. "Saving the Phenomena." *The Philosophical Review* 97 (1988): 303–52.

Borisov, Stanislav V., and Nina V. Podberezskaya. "X-ray Diffraction Analysis: A Brief History and Achievements of the First Century." *Journal of Structural Chemistry* 53 (2012): 1–3.

Borrelli, Ariana, and Janina Wellmann, eds. "Computer Simulations." *NTM Zeitschrift für Geschichte der Wissenschaften, Technik und Medizin* 27, no. 4, special issue (2019).

Bourdieu, Pierre. *The Logic of Practice*. Translated by Richard Nice. Cambridge: Polity Press, 1990. Originally published as *Le Sens pratique* (Paris: Editions Minuit, 1980).

Bourdieu, Pierre. *Pascalian Meditations*. Translated by Richard Nice. Stanford, CA: Stanford University Press, 2000. Originally published as *Méditations pascaliennes* (Paris: Seuil, 1997).

Bourdieu, Pierre, and Loïc J. D. Wacquant. *An Invitation to Reflexive Sociology*. Chicago: University of Chicago Press, 1992.

Brain, Robert. *The Pulse of Modernism: Physiological Aesthetics in Fin de Siècle Europe*. Spokane: University of Washington Press, 2015.

Brandt, Christina. *Metapher und Experiment: Von der Virusforschung zum genetischen Code*. Göttingen: Wallstein, 2004.

Bredekamp, Horst, Bernhard Buschendorf, Freia Hartung, and John Krois, eds. *Edgar Wind: Kunsthistoriker und Philosoph*. Berlin: Akademie Verlag, 1999.

Brock, Thomas D. *The Emergence of Bacterial Genetics*. New York: Cold Spring Harbor Laboratory Press, 1990.

Brunner, Reinhard. *Die Fragmentierung der Vernunft:. Rationalitätskritik im 20. Jahrhundert*. Frankfurt am Main: Campus, 1994.

Buchner, Eduard. "Alkoholische Gährung ohne Hefezellen." *Berichte der Deutschen Chemischen Gesellschaft* 30 (1897): 117–24.

Burian, Richard M. "How the Choice of Experimental Organism Matters: Epistemological Reflections on an Aspect of Experimental Practice." *Journal of the History of Biology* 26 (1993): 351–67.

Burri, Regula Valérie, and Joseph Dumit. "Social Studies of Scientific Imaging and Visualization." In *The Handbook of Science and Technology Studies*, ed. Edward J. Hackett, Olga Amsterdamska, Michael Lynch, and Judy Wajcman, 297–317. Cambridge, MA: MIT Press, 2008.

Campe, Rüdiger. "Kritzeleien im Sudelbuch: Zu Lichtenbergs Schreibverfahren." In *Über Kritzeln*, ed. Christian Driesen, Rea Köppel, Benjamin Meyer-Krahmer, and Eike Wittrock, 165–88. Zürich: Diaphanes, 2012.

Campe, Rüdiger. "'Unsere kleinen blinden Fertigkeiten.' Zur Entstehung des Wissens und zum Verfahren des Schreibens in Lichtenbergs Sudelbüchern." In *Lichtenberg-Jahrbuch* 2011, ed. Wolfgang Promies, Ulrich Joost, Alexander Neumann, and Heinrich Tuitje, 7–32. Heidelberg: Universitätsverlag Winter, 2012.

Campos, Luis A. *Radium and the Secret of Life*. Chicago: University of Chicago Press, 2015.

Canguilhem, Georges. *La Formation du concept de réflexe aux XVIIᵉ et XVIIIᵉ siècles*. Paris: Presses Universitaires de France, 1955.

Canguilhem, Georges. *Ideology and Rationality in the History of Life Sciences*. Translated by Arthur Goldhammer. Cambridge, MA: MIT Press, 1988. Originally published as *Idéologie et rationalité dans l'histoire des sciences de la vie* (Paris: Vrin, 1977).

Canguilhem, Georges. "The Object of the History of Sciences" (1968). Translated by Mary Tiles. In *Continental Philosophy of Science*, ed. Gary Gutting, 198–207. Oxford: Blackwell Publishing, 2005.

Canguilhem, Georges. "The Role of Analogies and Models in Biological Discovery." In *Scientific Change*, ed. Alistair C. Crombie, 507–20. London: Heinemann, 1963.

Carrel, Alexis, and Montrose T. Burrow. "Cultivation of Tissue In Vitro and Its Technique." *Journal of Experimental Medicine* 13 (1911): 387–96.

Cassirer, Ernst. *The Logic of the Cultural Sciences: Five Studies.* Translated by Steve G. Lofts. New Haven, CT: Yale University Press, 2000. Originally published as *Zur Logik der Kulturwissenschaften: Fünf Studien* (Göteborg: Wettergren & Kerber, 1942).

Cassirer, Ernst. *Philosophy of Symbolic Forms*, Vol. 1, *Language.* Translated by Ralph Manheim. New Haven, CT: Yale University Press, 1953. Originally published as *Philosophie der symbolischen Formen*, Vol. 1, *Die Sprache* (Berlin: Bruno Cassirer, 1923).

Cassirer, Ernst. *Philosophy of Symbolic Forms*, Vol. 2, *Mythical Thought.* Translated by Ralph Manheim. New Haven, CT: Yale University Press, 1955. Originally published as *Philosophie der symbolischen Formen*, Vol. 2, *Das mythische Denken* (Berlin: Bruno Cassirer, 1925).

Cassirer, Ernst. *Philosophy of Symbolic Forms*, Vol. 3, *Phenomenology of Knowledge.* Translated by Ralph Mannheim. New Haven, CT: Yale University Press, 1957. Originally published as *Philosophie der symbolischen Formen*, Vol. 3, *Phänomenologie der Erkenntnis* Berlin: Bruno Cassirer, 1929).

Cassirer, Ernst. *The Problem of Knowledge: Philosophy, Science and History since Hegel.* Translated by William H. Woglom and Charles W. Hendel. New Haven, CT: Yale University Press, 1950.

Cavaillès, Jean. "Les Deuxièmes Cours Universitaires de Davos." In *Die II. Davoser Hochschulkurse/Les Deuxièmes Cours Universitaires de Davos,* 65–81. Davos: Heintz, Neu, and Zahn, 1929.

Cavaillès, Jean. *On Logic and the Theory of Science.* Translated by Knox Peden and Robin Mackay. Cambridge, MA: MIT Press, 2021. Originally published as *Sur la logique et la théorie de la science* (Paris: Presses Universitaires de France, 1947).

de Chadarevian, Soraya. *Heredity under the Microscope: Chromosomes and the Study of the Human Genome.* Chicago: University of Chicago Press, 2021.

de Chadarevian, Soraya, and Nick Hopwood, eds. *Models: The Third Dimension of Science.* Stanford, CA: Stanford University Press, 2004.

de Chadarevian, Soraya, and Hans-Jörg Rheinberger, eds. "Disciplinary Histories and the History of Disciplines: The Challenge of Molecular Biology." *Studies in History and Philosophy of Biological and Biomedical Sciences* 40, no. 1, special issue (2009).

Chemla, Karine, and Evelyn Fox Keller, eds. *Cultures without Culturalism: The Making of Scientific Knowledge.* Durham, NC and London: Duke University Press, 2017.

Churchill, Frederick B., ed. "Special Section on Protozoology." *Journal of the History of Biology* 22 (1989): 185–323.

Cohen, Seymour. "The Biochemical Origins of Molecular Biology." *Trends in Biochemical Sciences* 9 (1984): 334–36.

Coopmans, Catelijne, Janet Vertesi, Michael Lynch, and Steve Woolgar, eds. *Representation in Scientific Practice Revisited.* Cambridge, MA: MIT Press, 2013.

Coveney, Peter, and Roger Highfield. *Frontiers of Complexity.* London: Faber & Faber, 1995.

Craver, Carl, and Lindley Darden, eds. "Mechanisms in Biology." *Studies in History and Philosophy of Biological and Biomedical Sciences* 36, no. 2, special issue (2005).

Creager, Angela N. H. *Life Atomic: A History of Radioisotopes in Science and Medicine.* Chicago: University of Chicago Press, 2013.

Creager, Angela N. H. *The Life of a Virus: Tobacco Mosaic Virus as an Experimental Model, 1930–1965.* Chicago: University of Chicago Press, 2002.

Creager Angela N. H. "Paradigms and Exemplars Meet Biomedicine." In *Kuhn's* Structure of Scientific Revolutions *at Fifty: Reflections on a Science Classic*, ed. Lorraine Daston and Robert Richards, 151–66. Chicago: University of Chicago Press, 2016.

Creager, Angela N. H. "Recipes for Recombining DNA: A History of Molecular Cloning: A Laboratory Manual." In "Learning by the Book: Manuals and Handbooks in the History of Knowledge," ed. Angela N. H. Creager, Mathias Grote, and Elaine Leong, *British Journal for the History of Science Themes* 5 (2020): 225–43.

Cremonini, Andreas, and Markus Klammer, eds. *Bild-Beispiele: Zu einer pikturalen Logik des Exemplarischen.* Munich: Fink, 2020.

Crombie, Alistair. *Styles of Scientific Thinking in the European Tradition: The History of Argument and Explanation Especially in the Mathematical and Biomedical Sciences and Arts.* London: Gerald Duckworth & Company, 1995.

Cuvier, Georges. *Essay on the Theory of the Earth.* Translated by Robert Kerr. New York: Kirk & Mercein, 1818. Originally published as *Discours préliminaire, Recherches sur les ossemens fossiles* (Paris: Deterville, 1812).

Dagognet, François. *Etienne Jules Marais: La Passion de la trace.* Paris: Hazan, 1987.

Dalgliesh, Charles E. "The Template Theory and the Role of Transpeptidation in Protein Biosynthesis." *Nature* 171 (1953): 1027–28.

Darbon, André. *Une Philosophie de l'expérience.* Paris: Presses Universitaires de France, 1946.

Daston, Lorraine, ed. *Biographies of Scientific Objects.* Chicago: University of Chicago Press, 2000.

Daston, Lorraine. "Taking Notes." *Isis* 95 (2004): 443–48.

Debru, Claude. *Georges Canguilhem, science et non-science.* Paris: Editions Rue d'Ulm, 2004.

Delaporte, François. *Figures de la médecine.* Paris: Les Editions du Cerf, 2009.

Delaporte, François, Bernard Devauchelle, and Emmanuel Fournier, eds. *Transplanter: Une approche transdisciplinaire: art, médecine, histoire et biologie.* Paris: Hermann, 2015.

Delbourgo, James, and Staffan Müller-Wille, eds. Focus Section "Listmania." *Isis* 103 (2012): 710–52.

Deleuze, Gilles. *Difference and Repetition.* Translated by Paul Patton. New York: Columbia University Press, 1994. Originally published as *Différence et repetition* (Paris: Presses Universitaires de France, 1968).

Derrida, Jacques. *Dissemination.* Translated by Barbara Johnson. Chicago: University of Chicago Press, 1981. Originally published as *La Dissémination* (Paris: Editions du Seuil, 1972).

Derrida, Jacques. *Edmund Husserl's Origin of Geometry: An Introduction.* Translated by John P. Leavey Jr. Lincoln, NE, and London: University of Nebraska Press, 1989. Originally published as *Edmund Husserl, L'Origine de la géométrie* (Paris: Presses Universitaires de France, 1962).

Derrida, Jacques. *Of Grammatology.* Translated by Chakravorty Spivak. Corrected edition. Baltimore: Johns Hopkins University Press, 1997. Originally published as *De la grammatologie* (Paris: Editions de Minuit, 1967).

Derrida, Jacques. "Signature Event Context." In *Limited Inc.*, trans. Samuel Weber, 1–23. Evanston, IL: Northwestern University Press, 1988. Originally published as "Signature événement contexte," in *Marges de la philosophie* (Paris: Editions de Minuit, 1972).

Desanti, Jean-Toussaint. *Phénoménologie et praxis.* Paris: Editions Sociales, 1970.

Desanti, Jean-Toussaint. *La Philosophie silencieuse, ou critique des philosophies de la science.* Paris: Editions du Seuil, 1975.

Diagrammatik-Reader. Grundlegende Texte aus Theorie und Geschichte. Berlin: De Gruyter, 2016.

Didi-Huberman, Georges. *Devant le temps: Histoire de l'art et anachronisme des images*. Paris: Editions de Minuit, 2000.

Didi-Huberman, Georges. *La Ressemblance par contact: Archéologie, anachronisme et modernité de l'empreinte*. Paris: Editions de Minuit, 2008.

Du Bois-Reymond, Emil. "Über die Lebenskraft: Aus der Vorrede zu den Untersuchungen über tierische Elektrizität (1848)." In *Vorträge über Philosophie und Gesellschaft*, ed. Siegfried Wollgast, 3–24. Hamburg: Felix Meiner, 1974.

Dunn, Leslie C. *A Short History of Genetics: The Development of Some of the Main Lines of Thought, 1864–1939*. New York: McGraw-Hill, 1965.

Dupré, John. *The Disorder of Things: Metaphysical Foundations of the Disunity of Science*. Cambridge, MA: Harvard University Press, 1993.

Ehrhardt, Caroline. "E uno plures? The Many Galois Theories (1832–1900)." In *Cultures without Culturalism: The Making of Scientific Knowledge*, ed. Karine Chemla and Evelyn Fox Keller, 327–51. Durham, NC and London: Duke University Press, 2017.

Ehrlich, Paul. "On Immunity, with Special Reference to Cell Life." *Proceedings of the Royal Society London* 66 (1900): 424–48.

Elkana, Yehuda. "Programmatic Attempt at an Anthropology of Knowledge." In *Sciences and Cultures: Anthropological and Historical Studies*, ed. Everett Mendelsohn and Yehuda Elkana, 1–76. Dordrecht: Reidel, 1981.

Emeis, Stefan, and Cornelia Lüdecke, eds. *From Beaufort to Bjerknes and Beyond: Critical Perspectives on Observing, Analyzing, and Predicting Weather and Climate*. Augsburg: Erwin Rauner Verlag (ERV), 2005.

Endersby, Jim. *A Guinea Pig's History of Biology*. Cambridge, MA: Harvard University Press, 2009.

Engelhardt, Wolf v. "'Der Versuch als Vermittler von Objekt und Subjekt': Goethes Aufsatz im Licht von Kants Vernunftkritik." *Athenäum: Jahrbuch für Romantik* 10 (2000): 9–28.

Enriques, Federigo. *Signification de l'histoire de la pensée scientifique*. Paris: Hermann, 1934.

Erdur, Onur. *Die epistemologischen Jahre: Philosophie und Biologie in Frankreich, 1960–1980*. Zurich: Chronos, 2018.

Espahangizi, Kijan. "From Topos to Oikos: The Standardization of Glass Containers as Epistemic Boundaries in Modern Laboratory Research (1850–1900)." *Science in Context* 28 (2015): 397–425.

Fernández-Millán, Pablo, Cédric Schelcher, Joseph Chihade, Benoît Masquida, Philippe Giegé, and Claude Sauter. "Transfer RNA: From Pioneering Crystallographic Studies to Contemporary tRNA Biology." *Archives of Biochemistry and Biophysics* 602 (2016): 95–105.

Fernbach, Auguste, and Louis Hubert. "De l'Influence des phosphates et de quelques autres matières minérales sur la diastase protéolytique du malt." *Comptes rendus des séances de l'Académie des Sciences* 131 (1900): 293–95.

Feyerabend, Paul. *Against Method*. 3rd edition. London and New York: Verso, 1993.

Feyerabend, Paul. *Wider den Methodenzwang: Skizze einer anarchistischen Erkenntnistheorie*. Frankfurt am Main: Suhrkamp, 1976.

Fischer, Emil. "Einfluss der Configuration auf die Wirkung der Enzyme." *Berichte der Deutschen Chemischen Gesellschaft* 27 (1894): 2985–93.

Fischer, Michael, and Paul Hoyningen-Huene. *Paradigmen: Facetten einer Begriffskarriere*. Frankfurt am Main and New York: Peter Lang, 1997.

Fischer, Philipp, Gabriele Gramelsberger, Christoph Hoffmann, Hans Hofman, Hans-Jörg Rheinberger, and Hannes Rickli. *Natures of Data*. Zurich and Berlin: Diaphanes, 2020.

Fleck, Ludwik. *Genesis and Development of a Scientific Fact*. Translated by Frederick Bradley and Thaddeus J. Trenn. Chicago: University of Chicago Press, 1979. Originally published as *Entstehung und Entwicklung einer wissenschaftlichen Tatsache: Einführung in die Lehre vom Denkstil und Denkkollektiv* (Basel: B. Schwabe, 1935).

Fleck, Ludwik. "Zur Krise der Wirklichkeit (1928)." In *Erfahrung und Tatsache: Gesammelte Aufsätze*, ed. Lothar Schäfer and Thomas Schnelle, 46–58. Frankfurt am Main: Suhrkamp, 1983.

Fortak, Heinz, and Paul Schlaak, eds. *100 Jahre Deutsche Meteorologische Gesellschaft in Berlin 1884–1984*. Berlin: Selbstverlag, 1984.

Foucault, Michel. *The Archaeology of Knowledge & The Discourse on Language*. Translated by A. M. Sheridan Smith and Rupert Swyer. New York: Pantheon Books, 1972. Originally published as *L'Archéologie du savoir* (Paris: Gallimard, 1969) and *L'Ordre du discours* (Paris: Gallimard, 1971).

Foucault, Michel. "Introduction." In *Le Rêve et l'existence*, by Ludwig Binswanger. Translated by Jacqueline Verdeaux. Brügge: Desclée de Brouwer, 1954. Originally published as *Traum und Existenz* (Zurich: H. Gersberger & Cie., 1930).

Foucault, Michel. *The Order of Things. An Archaeology of the Human Sciences*. New York: Pantheon Books, 1970. Originally published as *Les Mots et les choses* (Paris: Gallimard, 1966).

Fraassen, Bas van, and Isabelle Peschard, eds. *The Experimental Side of Modeling*. Minneapolis: University of Minnesota Press, 2018.

Frängsmyr, Tore. *History of Science in Sweden: The Growth of a Discipline, 1932–1982*. Uppsala: University of Uppsala, 1985.

Franklin, Allan, and Ephraim Fischbach. *The Rise and Fall of the Fifth Force: Discovery, Pursuit, and Justification in Modern Physics*. 2nd edition. Cham: Springer, 2016.

Frege, Gottlob. "On Sense and Reference." In *Translations from the Philosophical Writings of Gottlob Frege*, 3rd edition, ed. Peter Geach and Max Black, 56–78. Oxford: Blackwell, 1980. Originally published as "Über Sinn und Bedeutung," *Zeitschrift für Philosophie und philosophische Kritik* 100, NF (1892): 25–50.

Freudenthal, Gad, ed. *Etudes sur/Studies on Hélène Metzger*. Leiden: Brill, 1990.

Friedmann, Herbert C. "From Friedrich Wöhler's Urine to Eduard Buchner's Alcohol." In *New Beer in an Old Bottle: Eduard Buchner and the Growth of Biochemical Knowledge*, ed. Athel Cornish-Bowden, 67–122. València: Universitat de València, 1997.

Frigg, Roman. "Models and Fiction." *Synthese* 172 (2010): 251–68.

Fruton, Joseph S. *Molecules and Life: Historical Essays on the Interplay of Chemistry and Biology*. New York: Wiley, 1972.

Fruton, Joseph S. *Proteins, Enzymes, Genes: The Interplay of Chemistry and Biology*. New Haven, CT: Yale University Press, 1999.

Gachelin, Gabriel, ed. *Les Organismes modèles dans la recherche médicale*. Paris: Presses Universitaires de France, 2006.

Gadrois, Claude *Le Système du monde, selon les trois hypothèses, où conformément aux loix de la Mechanique l'on explique dans la supposition du mouvement de la Terre les apparences des astres [. . .]*. Paris: Guillaume Desprez, 1675.

Galison, Peter. *Image & Logic: A Material Culture of Microphysics*. Chicago: University of Chicago Press, 1997.

Galison, Peter, and David J. Stump, eds. *The Disunity of Science: Boundaries, Contexts, and Power*. Stanford, CA: Stanford University Press, 1996.

García Canclini, Néstor. *Hybrid Cultures*. Minneapolis: University of Minnesota Press, 1990.

García-Sancho, Miguel. *Biology, Computing, and the History of Molecular Sequencing: From Proteins to DNA, 1945–2000*. London: Palgrave Macmillan, 2012.

Garrido Wainer, Juan Manuel, Carla Fardella, and Juan Felipe Espinosa Cristia. "Arche-writing and Data-production in Theory-oriented Scientific Practice: The Case of Free-viewing as Experimental System to Test the Temporal Correlation Hypothesis." *History and Philosophy of the Life Sciences* 43 (2021): 70.

Gaudillière, Jean-Paul. *Inventer la biomédecine: La France, l'Amerique et la production des savoirs du vivant (1945–1965)*. Paris: Editions la Découverte, 2002.

Gaudillière, Jean-Paul, and Ilana Löwy, eds. *The Invisible Industrialist*. London: Palgrave Macmillan, 1998.

Gaudillière, Jean-Paul, and Hans-Jörg Rheinberger, eds. *The Mapping Cultures of Twentieth Century Genetics*, Vol. 2, *From Molecular Genetics to Genomics*. London: Routledge, 2004.

Gausemeier, Bernd. "Pedigrees of Madness: The Study of Heredity in Nineteenth and Early Twentieth Century Psychiatry." *History and Philosophy of the Life Sciences* 36 (2015): 467–83.

Gayon, Jean. *Gaston Bachelard: le rationalisme appliqué*. Paris: Centre National d'Enseignement à Distance/Presses Universitaires de France, 1994.

Gayon, Jean. "Les Organismes modèles en biologie et en médecine." In *Les Organismes modèles dans la recherche médicale*, ed. Gabriel Gachelin, 9–43. Paris: Presses Universitaires de France, 2006.

Geertz, Clifford. *The Interpretation of Cultures*. New York: Basic Books, 1973.

Geison, Gerald L., and Frederic L. Holmes, eds. *Research Schools: Historical Reappraisals* (*Osiris* 8). Chicago: University of Chicago Press, 1993.

Gelfert, Axel, ed. "Model-Based Representation in Scientific Practice." *Studies in History and Philosophy of Science* 42, no. 2, special issue (2011).

Ginzburg, Carlo. *The Cheese and the Worms: The Cosmos of a Sixteenth-Century Miller*. Translated by John Tedeschi and Anne C. Tedeschi. New York: Penguin Books, 1982. Originally published as *Il formaggio e i vermi* (Turin: Einaudi, 1976).

Ginzburg, Carlo. *Clues, Myths and the Historical Method*. Translated by John Tedeschi and Anne C. Tedeschi. Baltimore, MD: Johns Hopkins University Press, 1989. Originally published as *Miti emblemi spie. Morfologia e storia* (Turin: Einaudi, 1986).

Ginzburg, Carlo. "Microhistory." In *Threads and Traces: True False Fictive*, trans. Anne C. Tedeschi and John Tedeschi, 193–214. Berkeley, Los Angeles, London: University of California Press, 2012.

Godfrey-Smith, Peter. "The Strategy of Model-based Science." *Biology and Philosophy* 21 (2006): 725–40.

Goethe, Johann Wolfgang von. "The Experiment as Mediator of Object and Subject." Translated by Craig Holdrege. *Context* 24 (Fall 2010): 19–23. Originally published as "Der Versuch als Vermittler von Objekt und Subjekt" (Stuttgart and Tübingen: Cotta, 1823).

Goethe, Johann Wolfgang von. *Zur Naturwissenschaft überhaupt*, 2. Band, 1. Heft. Stuttgart and Tübingen: Cotta, 1823.

Goodman, Nelson. *Languages of Art: An Approach to a Theory of Symbols*. Indianapolis, IN and Cambridge: Hackett Publishing Company, 1976.

Goody, Jack. "What's In a List?" In *The Domestication of the Savage Mind*, 74–111. Cambridge: Cambridge University Press, 1977.

Gould, George M. *Illustrated Dictionary of Medicine, Biology, and Allied Sciences*. Philadelphia: P. Blakiston's Son & Co., 1894.

Gould, Stephen J. *Time's Arrow, Time's Cycle*. Cambridge, MA: Harvard University Press, 1987.

Gramelsberger, Gabriele, ed. *From Science to Computational Sciences: Studies in the History of Computing and Its Influence on Today's Sciences*. Zurich and Berlin: Diaphanes, 2011.

Grasshoff, Gerd, and Kärin Nickelsen. *Dokumente zur Entdeckung des Harnstoffzyklus*, Vol. 1, *Laborbuch Hans Krebs und Erstpublikationen*. Bern: Bern Studies in the History and Philosophy of Science, 2001.

Grasshoff, Gerd, Robert Casties, and Kärin Nickelsen. *Zur Theorie des Experiments: Untersuchungen am Beispiel der Entdeckung des Hanstoffzyklus*. Bern: Bern Studies in the History and Philosophy of Science, 2001.

Green, Sara. "When One Model Is Not Enough: Combining Epistemic Tools in Systems Biology." *Studies in History and Philosophy of Biological and Biomedical Sciences* 44 (2013): 170–80.

Grene, Marjorie. *The Knower and the Known*. Washington, DC: Center for Advanced Research in Phenomenology & University Press of America, 1984.

Griesemer, James, and Grant Yamashita. "Zeitmanagement bei Modellsystemen. Drei Beispiele aus der Evolutionsbiologie." In *Lebendige Zeit*, ed. Henning Schmidgen, 213–41. Berlin: Kadmos, 2005.

Grmek, Mirko D. "Introduction." In *Claude Bernard, Cahier de notes 1850–1865*, ed. Mirko D. Grmek, 11–25. Paris: Gallimard, 1965.

Grmek, Mirko D. *Raisonnement expérimental et recherches toxicologiques chez Claude Bernard*. Geneva and Paris: Librairie Droz, 1973.

Gros, François, Howard Hiatt, Walter Gilbert, Chuck G. Kurland, Robert W. Risebrough, and James D. Watson. "Unstable Ribonucleic Acid Revealed by Pulse Labelling of *E. coli*." *Nature* 190 (1961): 581–85.

Grote, Mathias. *Membranes to Molecular Machines: Active Matter and the Remaking of Life*. Chicago: University of Chicago Press, 2019.

Hacking, Ian. *Representing and Intervening: Introductory Topics in the Philosophy of Natural Science*. Cambridge: Cambridge University Press, 1983.

Hacking, Ian. "'Style' for Historians and Philosophers." *Studies in History and Philosophy of Science Part A* 23, no. 1 (1992): 1–20.

Harrison, Ross G. "The Outgrowth of the Nerve Fiber as a Mode of Protoplasmic Movement." *Journal of Experimental Zoology* 9 (1910): 787–846.

Hase, Albrecht. "Insekten." In *Methodik der wissenschaftlichen Biologie*, Vol. 2, ed. Tibor Péterfi, 265–89. Berlin: Springer, 1928.

Heering, Peter, Falk Rieß, and Christian Sichau, eds. *Im Labor der Physikgeschichte: Zur Untersuchung historischer Experimentalpraxis*. Oldenburg: BIS-Verlag, 2000.

Heider, Fritz. *Ding und Medium*. Edited by Dirk Baecker. Berlin: Kadmos, 2005. First published 1926.

Heisenberg, Werner. *Wandlungen in den Grundlagen der Naturwissenschaft: Zwei Vorträge*. Leipzig: S. Hirzel, 1935.

Hennig, Jochen. *Bildpraxis: Visuelle Strategien in der frühen Nanotechnologie*. Bielefeld: Transcript, 2011.

Hertz, Heinrich. *The Principles of Mechanics Presented in a New Form*. Translated by D. E. Jones and J. T. Walley. London: Macmillan and Company, 1899. Originally published as *Prinzipien der Mechanik in neuem Zusammenhange dargestellt* (Leipzig: J. A. Barth, 1894).

Hevesy, Georg v. "Historical Sketch of the Biological Application of Tracer Elements." *Cold Spring Harbor Symposia on Quantitative Biology* 13 (1948): 129–50.

Hildebrandt, Toni. "Postapokalyptisches Staunen: Ästhetik und Geschichtsbewusstsein im Naturvertrag." In *Staunen: Perspektiven eines Phänomens zwischen Natur und Kultur*, ed. Timo Kehren, Carolin Krahn, Georg Oswald, and Christoph Poetsch, 237–59. Paderborn: Fink, 2019.

Hilgartner, Stephen. "Biomolecular Databases: New Communication Regimes for Biology?" *Science Communication* 17 (1997): 506–22.

Hilgartner, Stephen. *Reordering Life: Knowledge and Control in the Genomics Revolution*. Cambridge, MA: MIT Press, 2017.

Hoagland, Mahlon. "On an Enzymatic Reaction between Amino Acids and Nucleic Acid and Its Possible Role in Protein Synthesis." *Recueil des travaux chimiques des Pays-Bas et de la Belgique* 77 (1958): 623–33.

Hoagland, Mahlon. *Toward the Habit of Truth: A Life in Science*. New York: W. W. Norton, 1990.

Hoffmann, Christoph. *Die Arbeit der Wissenschaften*. Zürich and Berlin: Diaphanes, 2013.

Hoffmann, Christoph. "Probe, Konservat, Modell—Präparat?" In *Präparat Bergsturz*, ed. Katharina Ammann and Priska Gisler, 61–69. Luzern-Poschiavo-Chur: Edizioni Periferia, Bündner Kunstmuseum, 2012.

Hoffmann, Christoph. *Schreiben im Forschen*. Tübingen: Mohr Siebeck, 2018.

Hoffmann, Christoph. *Unter Beobachtung: Naturforschung in der Zeit der Sinnesapparate*. Göttingen: Wallstein, 2006.

Hoffmann, Christoph, ed. *Daten sichern: Schreiben und Zeichnen als Verfahren der Aufzeichnung*. Zürich: Diaphanes, 2008.

Hoffmann, Christoph, and Peter Berz, eds. *Über Schall: Ernst Machs und Peter Salchers Geschossfotografien*. Göttingen: Wallstein, 2001.

Holbach, Paul Henri Thiry, Baron de. *Système de la nature ou des loix du monde physique & du monde moral*. London: M. M. Rey, 1770.

Holmberg, Christine. *The Politics of Vaccination: A Global History*. Manchester: Manchester University Press, 2017.

Holmes, Frederic L. *Claude Bernard and Animal Chemistry*. Cambridge, MA: Harvard University Press, 1974.

Holmes, Frederic L. *Hans Krebs: Architect of Intermediary Metabolism*. Oxford: Oxford University Press, 1993.

Holmes, Frederic L. *Investigative Pathways: Patterns and Stages in the Careers of Experimental Scientists*. New Haven, CT: Yale University Press, 2004.

Holmes, Frederic L. "Wissenschaftliches Schreiben und wissenschaftliches Entdecken." In *Schreiben als Kulturtechnik*, ed. Sandro Zanetti, 412–40. Frankfurt am Main: Suhrkamp, 2012.

Holmes, Frederic L., Jürgen Renn, and Hans-Jörg Rheinberger, eds. *Reworking the Bench: Research Notebooks in the History of Science*. Dordrecht: Kluwer, 2003.

Howlett, Peter, and Mary S. Morgan, eds. *How Well Do Facts Travel? The Dissemination of Reliable Knowledge*. Cambridge: Cambridge University Press, 2010.

Husserl, Edmund. *Méditations cartésiennes. Introduction à la phénoménologie*. Translated by Emmanuel Levinas and Gabrielle Peiffer. Paris: Armand Colin, 1931.

Husserl, Edmund. "The Origin of Geometry." In *Edmund Husserl's Origin of Geometry: An Introduction*. Edited and prefaced by Jacques Derrida. Translated by John P. Leavey Jr. Lincoln, NE, and London: University of Nebraska Press, 1989. Originally published as "Die Frage nach dem Ursprung der Geometrie als intentionalhistorisches Problem," *Research in Phenomenology* 1 (1939): 203–25.

Hyder, David. "Foucault, Cavaillès, and Husserl on the Historical Epistemology of the Sciences." *Perspectives on Science* 10 (2003): 107–29.

Ihde, Don. *Postphenomenology and Technoscience: The Peking University Lectures.* Albany: State University of New York Press, 2009.

Ihde, Don. *Experimental Phenomenology: Multistabilities.* 2nd edition. Albany: State University of New York Press, 2012.

Jacob, François. *The Logic of Life: A History of Heredity.* Translated by Betty E. Spillmann. New York: Pantheon Books, 1973. Originally published as *La Logique du vivant: Une histoire d'hérédité* (Paris: Gallimard, 1970).

Jacob, François. *Of Flies, Mice, and Men.* Translated by Giselle Weiss. Cambridge, MA and London: Harvard University Press, 1998. Originally published as *La Souris, la mouche, et l'homme* (Paris: Odile Jacob, 1997).

Jacob, François. *The Possible and the Actual.* Translated by Betty E. Spillmann. Seattle: University of Washington Press, 1982. Originally published as *Le Jeu des possibles: Essai sur la diversité du vivant* (Paris: Fayard, 1981).

Jacob, François. *The Statue Within: An Autobiography.* Translated by Franklin Philip. New York: Basic Books, 1988. Originally published as *La Statue intérieure* (Paris: Jacob, 1987).

Jansen, Sarah. "An American Insect in Imperial Germany: Visibility and Control in Making the Phylloxera in Germany, 1870–1914." *Science in Context* 13 (2000): 31–70.

Joerges, Bernward, and Terry Shinn, eds. *Instrumentation between Science, State, and Industry.* Dordrecht: Reidel, 2001.

Johannsen, Wilhelm. "The Genotype Conception of Heredity." *American Naturalist* 45 (1911): 129–59.

Kaltschmidt, Eberhard, and Heinz Günter Wittmann. "Ribosomal Proteins. XII. Number of Proteins in Small and Large Ribosomal Subunits of Escherichia coli as Determined by Two-dimensional Gel Electrophoresis." *Proceedings of the National Academy of Sciences of the United States of America* 67 (1970): 1276–82.

Kastenhofer, Karen. "Synthetic Biology as Understanding, Control, Construction, and Creation? Techno-epistemic and Socio-political Implications of Different Stances in Talking and Doing Technoscience." *Futures* 48 (2013): 13–22.

Kay, Lily E. *Who Wrote the Book of Life? A History of the Genetic Code.* Stanford, CA: Stanford University Press, 2000.

Keating, Peter, and Alberto Cambrosio. *Biomedical Platforms: Realigning the Normal and the Pathological in Late Twentieth-Century Medicine.* Cambridge, MA: MIT Press, 2003.

Kehren, Timo, Carolin Krahn, Georg Oswald, and Christoph Poetsch, eds. *Staunen: Perspektiven eines Phänomens zwischen Natur und Kultur.* Paderborn: Fink, 2019.

Keil, Maria. "Verwischte Grenzen zwischen Präparat und Modell: Epistemische Mustererkennung in medizinhistorischen Sammlungen." *Medizinhistorisches Journal* 55 (2020): 75–85.

Keller, Evelyn Fox. *The Century of the Gene.* Cambridge, MA: Harvard University Press, 2000.

Keller, Evelyn Fox. "Models Of and Models For: Theory and Practice in Contemporary Biology." *Philosophy of Science* 67, supplement (2000): S72–S86.

Keller, Evelyn Fox. "Models, Simulations, and Computer Experiments." In *The Philosophy of Scientific Experimentation*, ed. Hans Radder, 198–215. Pittsburgh: University of Pittsburgh Press, 2003.

Kellert, Stephen H., Helen E. Longino, and C. Kenneth Waters, eds. *Scientific Pluralism.* Minneapolis: University of Minnesota Press, 2006.

Kittler, Friedrich. *Discourse Networks 1800/1900*. Translated by Michael Metteer with Chris Cullens. Stanford, CA: Stanford University Press, 1990. Originally published as *Aufschreibesysteme 1800/1900* (Munich: Wilhelm Fink, 1985).

Klein, Ursula. *Experiments, Models, Paper Tools: Cultures of Organic Chemistry in the Nineteenth Century*. Stanford, CA: Stanford University Press, 2002.

Klein, Ursula. "Paper Tools in Experimental Cultures." *Studies in History and Philosophy of Science Part A* 32 (2001): 265–302.

Kleinschmidt, Albrecht K., Dimitrij Lang, Diether Jachters, and Rudolf K. Zahn. "Darstellung und Längenmessung des gesamten Desoxyribonucleinsäure-Inhaltes von T2-Bakteriophagen." *Biochimica et Biophyisca Acta* 61 (1962): 857–64.

Knorr Cetina, Karin. *Epistemic Cultures: How the Sciences Make Knowledge*. Cambridge, MA: Harvard University Press, 1999.

Kogge, Werner. *Experimentelle Begriffsforschung: Philosophische Interventionen am Beispiel von Code, Information und Skript in der Molekularbiologie*. Weilerswist: Velbrück, 2017.

Kohler, Robert E. "The Background to Eduard Buchner's Discovery of Cell-free Fermentation." *Journal of the History of Biology* 4 (1971): 35–61.

Kohler, Robert E. *From Medical Chemistry to Biochemistry: The Making of a Biomedical Discipline*. Cambridge: Cambridge University Press, 1982.

Kohler, Robert E. *Lords of the Fly: Drosophila Genetics and the Experimental Life*. Chicago: University of Chicago Press, 1994.

Koselleck, Reinhart. *Sediments of Time: On Possible Histories*. Translated by Sean Franzel and Stefan-Ludwig Hoffmann. Stanford, CA: Stanford University Press, 2018. Originally published as *Zeitschichten: Studien zur Historik* (Frankfurt am Main: Suhrkamp, 2000).

Koyré, Alexandre. *From the Closed World to the Open Universe*. Baltimore, MD: The Johns Hopkins University Press, 1957.

Krämer, Sybille. *Figuration, Anschauung, Erkenntnis: Grundlinien einer Diagrammatologie*. Berlin: Suhrkamp, 2016.

Krämer, Sybille, Werner Kogge, and Gernot Grube, eds. *Spur: Spurenlesen als Orientierungstechnik und Wissenskunst*. Frankfurt am Main: Suhrkamp, 2007.

Kubler, George. *The Shape of Time: Remarks on the History of Things*. New Haven, CT and London: Yale University Press, 1962.

Kühn, Alfred. "Grenzprobleme zwischen Vererbungsforschung und Chemie." *Berichte der Deutschen Chemischen Gesellschaft* 71 (1938): 107–14.

Kühn, Alfred. Letter to Ernst Caspari, December 27, 1946. In *Alfred Kühn (1885 bis 1968)— Lebensbilder in Briefen*, ed. Reinhard Mocek. Rangsdorf: Basilisken-Presse, 2020.

Kühn, Alfred. "Über eine Gen-Wirkkette der Pigmentbildung bei Insekten." *Nachrichten der Akademie der Wissenschaften in Göttingen, Mathematisch-Physikalische Klasse* (1941): 231–61.

Kühn, Alfred, and Karl Henke. *Genetische und entwicklungsphysiologische Untersuchungen an der Mehlmotte Ephestia kühniella Zeller, I–VI*. Abhandlungen der Gesellschaft der Wissenschaften zu Göttingen, Mathematisch-Physikalische Klasse, N.F. Bd. 15. Berlin: Weidmannsche Buchhandlung, 1929.

Kuhn, Thomas S. *The Structure of Scientific Revolutions*. 2nd edition, enlarged. Chicago: University of Chicago Press, 1970. First published 1962.

Lake, James A., David D. Sabatini, and Yoshiaki Nonomura. "Ribosome Structure as Studied by Electron Microscopy." In *Ribosomes*, ed. Masayasu Nomura, Alfred Tissières, and Pierre Lengyel, 543–57. New York: Cold Spring Harbor Laboratory Press, 1974.

Landecker, Hannah. "Building 'a New Type of Body in Which to Grow a Cell': Tissue Culture at the Rockefeller Institute, 1910–1914." In *Creating a Tradition of Biomedical Research: Contributions to the History of The Rockefeller University*, ed. Darwin H. Stapleton, 151–74. New York: The Rockefeller University Press, 2004.

Landecker, Hannah. *Culturing Life: How Cells Became Technologies*. Cambridge, MA: Harvard University Press, 2007.

Latour, Bruno. "Drawing Things Together." In *Representation in Scientific Practice*, ed. Michael Lynch and Steve Woolgar, 19–68. Cambridge, MA: MIT Press, 1990.

Latour, Bruno. *The Pasteurization of France*. Translated by Alan Sheridan and John Law. Cambridge, MA: Harvard University Press, 1993. Orignially published as *Les Microbes: Guerre et paix, suivi de Irréductions* (Paris: Métailié, 1984).

Latour, Bruno. "The 'Pedofil' of Boa Vista: A Photo-philosophical Montage." *Common Knowledge* 4 (1995): 145–87.

Latour, Bruno. *Science in Action: How to Follow Scientists and Engineers Through Society*. Cambridge, MA: Harvard University Press, 1987.

Latour, Bruno. *We Have Never Been Modern*. Translated by Catherine Porter. Cambridge, MA: Harvard University Press, 1993, Originally published as *Nous n'avons jamais été modernes: Essai d'anthropologie symétrique* (Paris: La Découverte, 1991).

Latour, Bruno, and Steve Woolgar. *Laboratory Life: The Construction of Scientific Facts*. 2nd edition. Princeton, NJ: Princeton University Press, 1986.

Lefèbvre, Henri. *Rhythmanalysis: Space, Time and Everyday Life*. Translated by Stuart Elden and Gerald Moore. London and New York: Continuum, 2004. Originally published as *Eléments de rythmanalyse: Introduction à la connaissance des rythmes* (Paris: Syllepse, 1992).

Lehninger, Albert. *Biochemistry*. New York: Worth Publishers, 1970.

Lenoir, Timothy, ed. *Instituting Science: The Cultural Production of Scientific Disciplines*. Stanford, CA: Stanford University Press, 1997.

Leonelli, Sabina. *Data-Centric Biology: A Philosophical Study*. Chicago: University of Chicago Press, 2016.

Leonelli, Sabina. "Documenting the Emergence of Bio-ontologies; or, Why Researching Bioinformatics Requires HPSSB." *History and Philosophy of the Life Sciences* 32 (2010): 105–26.

Lepère, Alexis. *Pratique raisonnée de la taille du pêcher*. 2nd edition. Paris: Bouchard-Huzard, 1846.

Lettkemann, Eric. *Stabile Interdisziplinarität: Eine Biografie der Elektronenmikroskopie aus historisch-soziologischer Perspektive*. Baden-Baden: Nomos, 2016.

Levi, Giovanni. *L'Eredità immateriale: Carriera di un esorcista ne Piemonte del Seicento*. Turin: Einaudi, 1985.

Levi, Giovanni. "On Microhistory." In *New Perspectives on Historical Writing*, 2nd edition, ed. Peter Burke, 97–119. University Park: The Pennsylvania State University Press, 2001.

Levins, Richard. "The Strategy of Model Building in Population Biology." *American Scientist* 54 (1966): 421–31.

Lévi-Strauss, Claude. *Introduction to a Science of Mythology I: The Raw and the Cooked*. Translated by John and Doreen Weightman. New York: Harper and Row, 1969. Originally published as *Mythologiques*, Vol. 1, *Le Cru et le cuit* (Paris: Plon, 1964).

Lévi-Strauss, Claude. *Structural Anthropology*. Translated by Claire Jacobson and Brooke Grundfest Schoepf. New York: Basic Books, 1963. Originally published as *Anthropologie structurale* (Paris: Plon, 1958).

Lévi-Strauss, Claude. *Wild Thought.* Translated by Jeffrey Mehlman and John Leavitt. Chicago: University of Chicago Press, 2021. Originally published as *La Pensée sauvage* (Paris: Plon, 1962).

Levy, Arno, and Adrian Currie. "Model Organisms Are Not Models." *British Journal for the Philosophy of Science* 66 (2015): 327–48.

Lichtenberg, Georg Christoph. *Sudelbücher: Schriften und Briefe*, Vols. 1 and 2. Edited by Wolfgang Promies. Munich: Hanser, 1967.

Lichtenthaler, Frieder W. "Hundert Jahre Schlüssel-Schloss-Prinzip: Was führte Emil Fischer zu dieser Analogie?" *Angewandte Chemie* 106 (1994): 2456–67.

Licoppe, Christian. *La Formation de la pratique scientifique: Le discours de l'expérience en France et en Angleterre* (1630–1820). Paris: La Découverte, 1996.

Linnaeus, Carolus. *Systema naturae.* Leiden: Johan Wilhelm de Groot, 1735. 12th edition, Stockholm, 1766/1767.

Loison, Laurent, and Michel Morange, eds. *L'Invention de la régulation génétique: Les Nobel 1965 (Jacob, Lwoff, Monod) et le modèle de l'opéron dans l'histoire de la biologie.* Paris: Editions Rue d'Ulm, 2017.

Lovejoy, Arthur O. *The Great Chain of Being: A Study of the History of an Idea.* Cambridge, MA: Harvard University Press, 1936.

Löwy, Ilana. "Variances in Meaning in Discovery Accounts: The Case of Contemporary Biology." *Historical Studies in the Physical and Biological Sciences* 21 (1990): 87–121.

Löwy, Ilana, ed. "Microscopic Slides: Reassessing a Neglected Historical Resource." *History and Philosophy of the Life Sciences* 35, no. 3, special issue (2013).

Lutz, Eckart, Vera Jerjen, and Christine Putzo. eds. *Diagramm und Text: Diagrammatische Strukturen und die Dynamisierung von Wissen und Erfahrung.* Wiesbaden: Reihert, 2014.

Lyell, Charles. *A Manual of Elementary Geology; or, The Ancient Changes on the Earth and Its Inhabitants as Illustrated by Geological Monuments.* London: John Murray, 1851.

Lynch, Michael. "The Production of Scientific Images: Vision and Re-vision in the History, Philosophy, and Sociology of Science." *Communication & Cognition* 31 (1998): 213–28.

Lynch, Michael, and Steve Woolgar, eds. *Representation in Scientific Practice.* Cambridge, MA: MIT Press, 1990.

Lyotard, Jean-François. *The Postmodern Condition: A Report on Knowledge.* Translated by Geoffrey Bennington and Brian Massumi. Minneapolis: University of Minnesota Press, 1984. Originally published as *La Condition postmoderne* (Paris: Éditions de Minuit, 1979).

Mach, Ernst. *Contributions to the Analysis of Sensations.* Translated by C. M. Williams. Chicago: The Open Court Publishing Company, 1897. Originally published as *Die Analyse der Empfindungen und das Verhältnis des Physischen zum Psychischen* (Jena: G. Fischer, 1886).

Mach, Ernst. *The Science of Mechanics.* Translated by Thomas J. McCormack. Chicago: The Open Court Publishing Company, 1892. Originally published as *Die Mechanik in ihrer Entwicklung* (Leipzig: F. A. Brockhaus, 1883).

Machamer, Peter, Lindley Darden, and Carl F. Craver "Thinking About Mechanisms." *Philosophy of Science* 67 (2000): 1–25.

Macho, Thomas, and Annette Wunschel, eds. *Science & Fiction: Über Gedankenexperimente in Wissenschaft, Philosophie und Literatur.* Frankfurt am Main: Fischer, 2004.

Mahr, Bernd. "Das Mögliche im Modell und die Vermeidung der Fiktion." In *Science & Fiction: Über Gedankenexperimente in Wissenschaft, Philosophie und Literatur*, ed. Thomas Macho and Annette Wunschel, 161–82. Frankfurt am Main: Fischer, 2004.

Mahr, Bernd. *Schriften zur Modellforschung*. Paderborn: Mentis Verlag, 2022.

Mahr, Bernd, and Reinhard Wendler. *Modelle als Akteure: Fallstudien*. KIT-Report 156, Technische Universität Berlin, 2009. https://www.flp.tu-berlin.de/fileadmin/fg53/KIT-Reports /r156.pdf.

Maienschein, Jane. *Transforming Traditions in American Biology, 1880–1915*. Baltimore, MD and London: Johns Hopkins University Press, 1991.

Mannheim, Karl. *Structures of Thinking*. Vol. 10 of *The Collected Works of Karl Mannheim*. Translated by Jeremy J. Shapiro and Sierry Webar Nicholsen. Edited and introduced by David Kettler, Volker Meja, and Niko Stehr. London: Routledge, 1991.

Marquéz, Viter, and Knud H. Nierhaus, "tRNA and Synthetases." In *Protein Synthesis and Ribosome Structure: Translating the Genome*, ed. Knud H. Nierhaus and Daniel N. Wilson, 145–67. Weinheim: Wiley-VCH, 2004.

Maupeu, Sarah, Kerstin Schankweiler, and Stefanie Stallschus, eds. *Im Maschenwerk der Kunstgeschichte: Eine Revision von George Kublers "The Shape of Time."* Berlin: Kadmos, 2014.

Mayer, Andreas. *The Science of Walking: Investigations into Locomotion in the Long Nineteenth Century*. Translated by Robin Blanton and Tilman Skowroneck. Chicago: University of Chicago Press, 2020. Originally published as *Wissenschaft vom Gehen: Die Erforschung der Bewegung im 19. Jahrhundert* (Frankfurt am Main: Fischer, 2013).

Mayr, Ernst. *The Growth of Biological Thought: Diversity, Evolution, and Inheritance*. Cambridge, MA: The Belknap Press of Harvard University Press, 1982.

McDonald, Christine, ed. *The Ear of the Other: Texts and Discussions with Jacques Derrida*. Lincoln, NE, and London: University of Nebraska Press, 1988.

McGovern, Michael F. "Genes Go Digital: *Mendelian Inheritance in Man* and the Genealogy of Electronic Publishing in Biomedicine." *The British Journal for the History of Science* 54, no. 2 (2021): 1–19.

Medawar, Peter B. *The Art of the Soluble*. London: Methuen, 1967.

Medick, Hans. *Weben und Überleben in Laichingen 1650–1900: Lokalgeschichte als allgemeine Geschichte*. Göttingen: Vandenhoeck & Ruprecht, 1996.

Mendelsohn, Andrew. "'Like All that Lives': Biology, Medicine and Bacteria in the Age of Pasteur and Koch." *History and Philosophy of the Life Sciences* 24 (2002): 3–36.

Mengaldo, Elisabetta. *Zwischen Naturlehre und Rhetorik: Kleine Formen des Wissens in Lichtenbergs Sudelbüchern*. Göttingen: Wallstein, 2021.

Mertens, Rebecca. *The Construction of Analogy-Based Research Programs: The Lock-and-Key Analogy in 20th Century Biochemistry*. Bielefeld: Transcript, 2019.

Merton, Robert K. *Social Theory and Social Structure*. New York: The Free Press, 1968. First published 1957.

Merton, Robert K. "Three Fragments from a Sociologist's Notebooks: Establishing the Phenomenon, Specified Ignorance, and Strategic Research Materials." *Annual Review of Sociology* 13 (1987): 1–28.

Merton, Robert K., and Elinor Barber. *The Travels and Adventures of Serendipity: A Study in Sociological Semantic and the Sociology of Science*. Princeton, NJ: Princeton University Press, 2006.

Metzger, Hélène. *La Méthode philosophique en histoire des sciences: Textes 1914–1939*. Edited by Gad Freudenthal. Paris: Fayard, 1987.

Michaelis, Leonor, and Heinrich Davidsohn. "Die Abhängigkeit der Trypsinwirkung von der Wasserstoffionenkonzentration." *Biochemische Zeitschrift* 36 (1911): 280–90.

Michaelis, Leonor, and Maud Menten. "Die Kinetik der Invertinwirkung." *Biochemische Zeitschrift* 49 (1913): 333–69.

Morange, Michel. *The Black Box of Biology: A History of the Molecular Revolution*. Translated by Matthew Cobb. Cambridge, MA: Harvard University Press, 2020.

Morgan, Mary S. *The World in the Model: How Economists Work and Think*. Cambridge: Cambridge University Press, 2012.

Morgan, Mary S., and Margaret Morrison, eds. *Models as Mediators: Perspectives on Natural and Social Science*. Cambridge: Cambridge University Press, 1999.

Morgan, Mary S., and Norton Wise, eds. "Narrative in Science." *Studies in History and Philosophy of Science Part A* 62, special issue (2017): 1–98.

Morrison, Margaret. "Models as Autonomous Agents." In *Models as Mediators: Perspectives on Natural and Social Science*, ed. Mary S. Morgan and Margaret Morrison, 38–65. Cambridge: Cambridge University Press, 1999.

Morrison, Margaret. *Reconstructing Reality: Models, Mathematics, and Simulations*. Oxford and New York: Oxford University Press, 2015.

Moulin, Anne-Marie. *L'Aventure de la vaccination*. Paris: Fayard, 1996.

Müller, Falk. *Jenseits des Lichts: Siemens, AEG und die Anfänge der Elektronenmikroskopie in Deutschland*. Göttingen: Wallstein, 2021.

Müller, Johannes. "Jahresbericht über die Fortschritte der anatomisch-physiologischen Wissenschaften im Jahre 1833." *Archiv für Anatomie, Physiologie und wissenschaftliche Medizin* (1834): 1–201.

Müller, Olaf. *Mehr Licht: Goethe und Newton im Streit um die Farben*. Frankfurt am Main: Fischer, 2015.

Müller-Wille, Staffan and Isabelle Charmantier, "Lists as Research Technologies." *Isis* 103 (2012): 743–52.

Müller-Wille, Staffan, and Hans-Jörg Rheinberger. *A Cultural History of Heredity*. Chicago: University of Chicago Press, 2012. Originally published as *Vererbung: Geschichte und Kultur eines biologischen Konzepts* (Frankfurt am Main: Fischer Taschenbuch Verlag, 2009).

Nersessian, Nancy. "Engineering Devices: The Culture of Physical Simulation Modeling in Biomedical Engineering." In *Cultures without Culturalism: The Making of Scientific Knowledge*, ed. Karine Chemla and Evelyn Fox Keller, 117–44. Durham, NC and London: Duke University Press, 2017.

Nicholson, Daniel J. "Is the Cell Really a Machine?" *Journal of Theoretical Biology* 477 (2019): 108–26.

Nickelsen, Kärin, and Gerd Grasshoff. "Concepts from the Bench: Krebs and the Urea Cycle." In *Going Amiss in Experimental Research*, ed. Giora Hon, Jutta Schickore, and Friedrich Steinle, 91–117. Boston: Springer, 2009.

Niehaus, Michael, and Hans-Walter Schmidt-Hannisa, eds. *Das Protokoll: Kulturelle Funktionen einer Textsorte*. Frankfurt am Main: Peter Lang, 2005.

Nierhaus, Knud H. "The Elongation Cycle." In *Protein Synthesis and Ribosome Structure: Translating the Genome*, ed. Knud H. Nierhaus and Daniel Wilson, 323–66. Weinheim: Wiley-VCH, 2004.

Nomura, Masayasu, and William A. Held. "Reconstitution of Ribosomes: Studies of Ribosome Structure, Function and Assembly." In *Ribosomes*, ed. Masayasu Nomura, Alfred Tissières, and Pierre Lengyel, 193–223. New York: Cold Spring Harbor Laboratory Press, 1974.

Nowotny, Helga. *Time: The Modern and Postmodern Experience*. Translated by Neville Plaice. Cambridge: Polity Press, 1996. Originally published as *Eigenzeit. Entstehung und Entwicklung eines Zeitgefühls* (Frankfurt am Main: Suhrkamp, 1989).

Numbers, Ronald L., ed. *Newton's Apple and Other Myths about Science*. Cambridge, MA: Harvard University Press, 2015.

Oakes, Melanie, Eric Henderson, Andrew Scheinman, Michael W. Clark, and James A. Lake. "Ribosome Structure, Function, and Evolution: Mapping Ribosomal RNA, Proteins, and Functional Sites in Three Dimensions." In *Structure, Function, and Genetics of Ribosomes*, ed. Boyd Hardesty and Gisela Kramer, 47–67. New York: Springer, 1986.

Olby, Robert. *Origins of Mendelism*. New York: Schocken Books, 1967.

Olesen, Søren Gosvig. *Wissen und Phänomen: Eine Untersuchung der ontologischen Klärung der Wissenschaften bei Edmund Husserl, Alexandre Koyré und Gaston Bachelard*. Würzburg: Königshausen & Neumann, 1997.

O'Malley, Maureen A. "Making Knowledge in Synthetic Biology: Design Meets Kludge." *Biological Theory* 4 (2009): 378–89.

Palade, George E. "Intracellular Distribution of Acid Phosphatase in Rat Liver Cells." *Archives of Biochemistry* 30 (1951): 144–58.

Palade, George E. "A Small Particulate Component of the Cytoplasm." *Journal of Biophysical and Biochemical Cytology* 1 (1955): 59–68.

Palló, Gábor. "Isotope Research Before Isotopy: George Hevesy's Early Radioactivity Research in the Hungarian Context." *Dynamis* 29 (2009): 167–89.

Parnes, Ohad, Ulrike Vedder, and Stefan Willer. *Generation: Eine Geschichte der Wissenschaft und der Kultur*. Frankfurt am Main: Suhrkamp, 2008.

Pasztory, Esther. *Thinking with Things: Toward a New Vision of Art*. Austin: University of Texas Press, 2005.

Peirce, Charles Sanders. "Abduction and Induction." In *The Philosophical Writings of Peirce*, ed. Justus Buchler, 150–56. New York: Dover Publications, 1955.

Peirce, Charles Sanders. "Logic as Semiotic: The Theory of Signs." In *The Philosophical Writings of Peirce*, ed. Justus Buchler, 98–119. New York: Dover Publications, 1955.

Pickering, Andrew. *The Mangle of Practice: Time, Agency, and Science*. Chicago: University of Chicago Press, 1995.

Pickering, Andrew, ed. *Science as Practice and Culture*. Chicago: University of Chicago Press, 1992.

Pickstone, John V. "A Brief Introduction to Ways of Knowing and Ways of Working." *History of Science* 49 (2011): 235–45.

Pickstone, John V. *Ways of Knowing: A New History of Science, Technology and Medicine*. Manchester: Manchester University Press, 2000.

Pinheiro dos Santos, Lúcio Alberto. *La Rythmanalyse*. Rio de Janeiro: Société de Psychologie et de Philosophie, 1931.

Pomian, Krzysztof. *L'Ordre du temps*. Paris: Gallimard, 1984.

Polanyi, Michael. *Duke Lectures* (1964). Microfilm, Library Photographic Service, University of California, Berkeley, 1965.

Polanyi, Michael. "The Logic of Tacit Inference." In *Knowing and Being: Essays by Michael Polanyi*, ed. Marjorie Grene, 138–58. Chicago: University of Chicago Press, 1969. First published in *Philosophy* 41, no. 155 (1966).

Polanyi, Michael. "The Unaccountable Element in Science." In *Knowing and Being: Essays by Michael Polanyi*, ed. Marjorie Grene, 105–20. Chicago: University of Chicago Press, 1969. First published in *Philosophy* 37, no. 139 (1962).

Popper, Karl. *The Logic of Scientific Discovery*. New York: Basic Books, 1959. Originally published as *Logik der Forschung* (Vienna: J. Springer, 1935).

Porter, Theodore M. *Genetics in the Madhouse: The Unknown History of Human Heredity*. Princeton, NJ: Princeton University Press, 2018.

Prescott, David M. "Cellular Sites of RNA." *Progress in Nucleic Acid Research and Molecular Biology* 3 (1964): 33–57.

Rabinow, Paul. *Making PCR: A Story of Biotechnology*. Chicago: University of Chicago Press, 1996.

Rader, Karen. *Making Mice: Standardizing Animals for American Biomedical Research*. Princeton, NJ: Princeton University Press, 2004.

Raible, Wolfgang. *Sprachliche Texte—Genetische Texte: Sprachwissenschaft und molekulare Genetik*. Heidelberg: Winter, 1993.

Ramakrishnan, Venkatraman, Malcolm S. Capel, William M. Clemons Jr., Joanna L. C. May, and Brian T. Wimberly. "Progress Toward the Crystal Structure of a Bacterial 30S Ribosomal Subunit." In *The Ribosome: Structure, Function, Antibiotics, and Cellular Interactions*, ed. Roger Garrett, Stephen R. Douthwaite, Anders Liljas, Alistair T. Matheson, Peter B. Moore, and Harry F. Noller, 3–9. Washington, DC: ASM Press, 2000.

Rasmussen, Nicolas. *Gene Jockeys: Life Science and the Rise of Biotech Enterprise*. Baltimore, MD: Johns Hopkins University Press, 2014.

Rasmussen, Nicolas. *Picture Control: The Electron Microscope and the Transformation of Biology in America, 1940–1960*. Stanford, CA: Stanford University Press, 1997.

Reichenbach, Hans. *Experience and Prediction: An Analysis of the Foundations of the Structure of Knowledge*. Chicago: University of Chicago Press, 1938.

Reinhardt, Carsten. *Shifting and Rearranging: Physical Methods and the Transformation of Modern Chemistry*. Sagamore Beach, MA: Science History Publications/USA, 2006.

Reinhold, Sabine, and Kerstin P. Hofmann, eds. "Zeichen der Zeit: Archäologische Perspektiven auf Zeiterfahrung, Zeitpraktiken und Zeitkonzepte." *Forum Kritische Archäologie* 3, thematic issue (2014).

Reisch, George A. *The Politics of Paradigms: Thomas S. Kuhn, James B. Conant, and the Cold War "Struggle for Men's Minds."* Albany: State University of New York Press, 2019.

Reiß, Christian. *Der Axolotl: Ein Labortier im Heimaquarium, 1864–1914*. Göttingen: Wallstein, 2020.

Renn, Jürgen. *The Genesis of General Relativity*. Vols. 1 and 2 of *Einstein's Zurich Notebooks*. Dordrecht: Springer, 2007.

Rentetzi, Maria. *Trafficking Materials and Gendered Experimental Practices: Radium Research in Early 20th Century Vienna*. New York: Columbia University Press, 2008.

Rheinberger, Hans-Jörg. "Cytoplasmic Particles in Brussels (Jean Brachet, Hubert Chantrenne, Raymond Jeener) and at Rockefeller (Albert Claude), 1935–1955." *History and Philosophy of the Life Sciences* 19 (1997): 47–67.

Rheinberger, Hans-Jörg. *An Epistemology of the Concrete: Twentieth Century Histories of Life*. Durham, NC and London: Duke University Press, 2010b.

Rheinberger, Hans-Jörg. *Experiment, Differenz, Schrift*. Marburg: Basiliskenpresse, 1992.

Rheinberger, Hans-Jörg. "From the 'Originary Phenomenon' to the 'System of Pelagic Fishery': Johannes Müller (1801–1858) and the Relation Between Physiology and Philosophy." In *From Physico-Theology to Bio-Technology: Essays in the Social and Cultural History of Biosciences. A Festschrift for Mikulas Teich*, ed. Kurt Bayertz and Roy Porter, 133–52. Amsterdam: Rodopi, 1998.

Rheinberger, Hans-Jörg. *The Hand of the Engraver: Albert Flocon Meets Gaston Bachelard*. Translated by Kate Sturge. Albany: State University of New York Press, 2018. Originally published as *Der Kupferstecher und der Philosoph* (Zurich and Berlin: Diaphanes, 2016).

Rheinberger, Hans-Jörg. *On Historicizing Epistemology: An Essay.* Stanford, CA: Stanford University Press, 2010a.

Rheinberger, Hans-Jörg. "A History of Protein Synthesis and Ribosome Research." In *Protein Synthesis and Ribosome Structure: Translating the Genome,* ed. Knud H. Nierhaus and Daniel N. Wilson, 1–51. Weinheim: Wiley-VCH, 2004.

Rheinberger, Hans-Jörg. "Kurze Geschichte der Molekularbiologie." In *Geschichte der Biologie,* ed. Ilse Jahn, 642–63. Jena: Gustav Fischer, 1998.

Rheinberger, Hans-Jörg. "Vom Mikrosom zum Ribosom: 'Strategien' der 'Repräsentation' 1935– 1955." In *Die Experimentalisierung des Lebens: Experimentalsysteme in den biologischen Wissenschaften 1850/1950,* ed. Hans-Jörg Rheinberger and Michael Hagner, 162–87. Berlin: Akademie Verlag, 1993.

Rheinberger, Hans-Jörg. *Natur und Kultur im Spiegel des Wissens.* Heidelberg: Winter, 2015.

Rheinberger, Hans-Jörg. *Toward a History of Epistemic Things: Synthesizing Proteins in the Test Tube.* Stanford, CA: Stanford University Press, 1997.

Rheinberger, Hans-Jörg, and Jean-Paul Gaudillière. *The Mapping Cultures of Twentieth Century Genetics,* Vol. 1, *Classical Genetic Research and Its Legacy.* London: Routledge, 2004.

Rheinberger, Hans-Jörg, and Staffan Müller-Wille. *The Gene: From Genetics to Postgenomics.* Chicago: University of Chicago Press, 2017.

Ricœur, Paul. "Narrative Time." *Critical Inquiry* 7 (1980): 169–90.

Ricœur, Paul. "Symbole et temporalité." *Archivio di Filosofía* 1–2 (1963): 5–41.

Ricœur, Paul. *Time and Narrative,* Vol. 1. Translated by Kathleen McLaughlin and David Pellauer. Chicago: University of Chicago Press, 1984. Originally published as *Temps et récit,* Vol. 1 (Paris: Editions du Seuil, 1983).

Robinson, Keith Alan. *Michel Foucault and the Freedom of Thought.* New York: E. Mellen Press, 2001.

Rotman, Brian. *Signifying Nothing: The Semiotics of Zero.* New York: St. Martin's Press, 1987.

Rudwick, Martin. *Georges Cuvier, Fossil Bones, and Geological Catastrophes.* Chicago: University of Chicago Press, 1997.

Rudwick, Martin. *The Meaning of Fossils: Episodes in the History of Palaeontology.* London: MacDonald, 1972.

Rudwick, Martin. *Scenes from Deep Time: Early Pictorial Images of the Prehistoric World.* Chicago: University of Chicago Press, 1992.

Sabean, David Warren. *Property, Production and Family in Neckarhausen, 1700–1870.* Cambridge: Cambridge University Press, 1990.

Sachs-Hombach, Klaus, ed. *Bildtheorien: Anthropologische und kulturelle Grundlagen des Visualistic Turn.* Frankfurt am Main: Suhrkamp, 2009.

Sambrook, Joseph, Edward F. Fritsch, and Tom Maniatis. *Molecular Cloning: A Laboratory Manual.* Cold Spring Harbor, NY: Cold Spring Harbor Laboratory Press, 1982.

Sanger, Frederick, Steve Nicklen, and Alan R. Coulson. "DNA Sequencing with Chainterminating Inhibitors." *Proceedings of the National Academy of Sciences of the United States of America* 74 (1977): 5463–67.

Schickore, Jutta. *The Microscope and the Eye: A History of Reflections, 1740–1870.* Chicago: University of Chicago Press, 2007.

Schickore, Jutta, and Friedrich Steinle, eds. *Revisiting Discovery and Justification: Historical and Philosophical Perspectives on the Context Distinction.* Dordrecht: Springer, 2006.

Schlich, Thomas. *The Origins of Organ Transplantation: Surgery and Laboratory Science, 1880– 1930.* Rochester, NY: University of Rochester Press, 2010.

Schmid-Burkhardt, Astrit. *Die Kunst der Diagrammatik: Perspektiven eines neuen bildwissenschaftlichen Paradigmas*. Bielefeld: Transcript, 2017.

Schmidgen, Henning. *Forschungsmaschinen: Experimente zwischen Wissenschaft und Kunst*. Berlin: Matthes & Seitz, 2017.

Schmidgen, Henning. *The Helmholtz-Curves: Tracing Lost Time*. Translated by Nils F. Schott. New York: Fordham University Press, 2014a. Originally published as *Die Helmholtz-Kurven: Auf der Spur der verlorenen Zeit* (Berlin: Merve, 2009).

Schmidgen, Henning. *Hirn und Zeit: Die Geschichte eines Experiments*. Berlin: Matthes & Seitz, 2014b.

Schmidgen, Henning. "The Life of Concepts: Georges Canguilhem and the History of Science." *History and Philosophy of the Life Sciences* 36 (2014c): 232–53.

Schmidgen, Henning, ed. *Lebendige Zeit: Wissenskulturen im Werden*. Berlin: Kadmos, 2005.

Scholthof, Karen-Beth G., John G. Shaw, and Milton Zaitlin, eds. *Tobacco Mosaic Virus: One Hundred Years of Contributions to Virology*. St. Paul, MN: APS Press, 1999.

Schulze, Winfried, ed. *Sozialgeschichte, Alltagsgeschichte, Mikro-Historie: Eine Diskussion*. Göttingen: Vandenhoeck & Ruprecht, 1994.

Schwab, Michael, and Hans-Jörg Rheinberger. "Forming and Being Informed." A Conversation. In *Experimental Systems: Future Knowledge in Artistic Research*, ed. Michael Schwab, 198–219. Leuven: Leuven University Press, 2013.

Sepkoski, David. *Rereading the Fossil Record: The Growth of Paleobiology as an Evolutionary Discipline*. Chicago: University of Chicago Press, 2012.

Serres, Michel. *Conversations on Science, Culture, and Time: Michel Serres Interviewed by Bruno Latour*. Translated by Roxanne Lapidus. Ann Arbor: University of Michigan Press, 1995. Originally published as *Eclaircissements: Cinq entretiens avec Bruno Latour* (Paris: F. Bourin, 1992).

Serres, Michel. "Introduction." In *A History of Scientific Thought: Elements of a History of Science*, ed. Michel Serres. Oxford: Blackwell, 1995. Originally published as *Eléments d'histoire des sciences* (Paris: Bordas, 1989).

Serres, Michel. *The Natural Contract*. Translated by Elizabeth MacArthur and William Paulson. Ann Arbor: University of Michigan Press, 1995. Originally published as *Le Contrat naturel* (Paris: F. Bourin, 1990).

Shinn, Terry. *Research-Technology and Cultural Change: Instrumentation, Genericity, Transversality*. Oxford: Bardwell Press, 2007.

Sibum, Otto. "Reworking the Mechanical Value of Heat: Instruments of Precision and Gestures of Accuracy in Early Victorian England." *Studies in History and Philosophy of Science Part A* 26 (1995): 73–106.

Sismondo, Sergio. "Models, Simulations, and Their Objects." *Science in Context* 12 (1999): 247–60.

Sober, Elliott. "Independent Evidence about a Common Cause." *Philosophy of Science* 56 (1989): 275–87.

Sommer, Marianne. *Bones and Ochre: The Curious Afterlife of the Red Lady of Paviland*. Cambridge, MA: Harvard University Press, 2007.

Sørensen, Søren P. L. "Enzymstudien. II. Mitteilung. Über die Messung und die Bedeutung der Wasserstoffionenkonzentration bei enzymatischen Prozessen." *Biochemische Zeitschrift* 21 (1909): 131–304.

Spielmann, Yvonne. *Hybridkultur*. Frankfurt am Main: Suhrkamp, 2009.

Spirin, Alexander S. "A Model of the Functioning Ribosome: Locking and Unlocking of the Ribo-some Subparticles." *Cold Spring Harbor Symposia on Quantitative Biology* 34 (1969): 197–207.

Stabrey, Undine. *Archäologische Untersuchungen: Über Temporalität und Dinge.* Bielefeld: Tran-script, 2017.

Stamhuis, Ida H., Teun Koetsier, Cornelis de Pater, and Albert van Helden, eds. *The Changing Images of Unity and Disunity in the Philosophy of Science.* Dordrecht: Springer, 2002.

Star, Susan L., and James R. Griesemer. "Institutional Ecology, 'Translations' and Boundary Ob-jects: Amateurs and Professionals in Berkeley's Museum of Vertebrate Zoology, 1907–1939." *Social Studies of Science* 19 (1989): 387–420.

Stegenga, Jacob, and Tarun Menon. "Robustness and Independent Evidence." *Philosophy of Sci-ence* 84 (2017): 424–35.

Steinle, Friedrich. "Concepts, Facts, and Sedimentation in Experimental Science." In *Science and the Life-World: Essays on Husserl's Crisis of European Sciences*, ed. David Hyder and Hans-Jörg Rheinberger, 199–214. Stanford, CA: Stanford University Press, 2010.

Steinle, Friedrich. *Exploratory Experiments: Ampère, Faraday, and the Origins of Electrodynam-ics.* Translated by Alex Levine. Pittsburgh: University of Pittsburgh Press, 2016. Originally published as *Explorative Experimente: Ampère, Faraday und die Ursprünge der Elektrody-namik* (Stuttgart: Steiner, 2005).

Steinle, Friedrich. "Das Nächste ans Nächste reihen: Goethe, Newton und das Experiment." *Philosophia naturalis: Archiv für Naturphilosophie und die philosophischen Grenzgebiete der exakten Wissenschaften und Wissenschaftsgeschichte* 39 (2002): 141–72.

Stengers, Isabelle. "La Propagation des concepts." In *D'Une science à l'autre: Des concepts no-mades*, ed. Isabelle Stengers, 9–26. Paris: Editions du Seuil, 1987.

Stengers, Isabelle, ed. *D'Une science à l'autre: Des concepts nomades.* Paris: Editions du Seuil, 1987.

Stent, Gunther S. *Molecular Biology of Bacterial Viruses.* San Francisco: Freeman, 1963.

Stichweh, Rudolf. *Zur Entstehung des modernen Systems wissenschaftlicher Disziplinen: Physik in Deutschland 1740–1890.* Frankfurt am Main: Suhrkamp, 1984.

Stjernfeldt, Frederick. *Diagrammatology: An Investigation on the Borderlines of Phenomenology, Ontology, and Semiotics.* Dordrecht: Springer, 2007.

Strasser, Bruno. *La Fabrique d'une nouvelle science: La biologie moléculaire à l'âge atomique.* Flor-ence: Olschki, 2006.

Strasser, Bruno. *Collecting Experiments: Making Big Data Biology.* Chicago: University of Chi-cago Press, 2019.

Stubbe, Hans. *Kurze Geschichte der Genetik.* Jena: Fischer, 1963.

Suarez, Mauricio, ed. *Fictions in Science: Philosophical Essays on Modeling and Idealization.* Lon-don: Routledge, 2009.

Suarez Díaz, Edna. "Variation, Differential Reproduction and Oscillation: The Evolution of Nucleic Acid Hybridization." *History and Philosophy of the Life Sciences* 35 (2013): 39–44.

Suddath, Fred L., Gary Joseph Quigley, Alexander McPherson, Darryl Sneden, Jung Ja Park Kim, Sung Hou Kim, and Alexander Rich. "Three-dimensional Structure of Phenylalanine Transfer RNA at 3.0 Å Resolution." *Nature* 248 (1974): 20–24.

Summers, William C. "From Culture as Organism to Organism as Cell: Historical Origin of Bacterial Genetics." *Journal of the History of Biology* 24 (1991): 171–90.

Talcott, Samuel. *Georges Canguilhem and the Problem of Error.* London: Palgrave Macmillan, 2019.

Tarski, Alfred. *Introduction to Logic and the Methodology of Deductive Sciences.* Oxford: Oxford University Press, 1941. Originally published as *Einführung in die mathematische Logik und in die Methodologie der Mathematik* (Vienna: Springer, 1937).

Tengelyi, László. *Welt und Unendlichkeit: Zum Problem phänomenologischer Metaphysik.* Freiburg and Munich: Karl Alber, 2014.

Traut, Robert R., Ronald L. Heimark, Tung-Tien Sun, John W. B. Hershey, and Alex Bollen. "Protein Topography of Ribosomal Subunit from *Escherichia coli.*" In *Ribosomes,* ed. Masayasu Nomura, Alfred Tissières, and Pierre Lengyel, 271–308. New York: Cold Spring Harbor Laboratory Press, 1974.

Tzara, Tristan. *Grains et issues.* Edited by Henri Béhar. Paris: Garnier-Flammarion, 1981. First published 1935.

Unwin, P. Nigel T., and Carlo Taddei. "Packing of Ribosomes in Crystals from the Lizard *Lacerta sicula.*" *Journal of Molecular Biology* 114 (1977): 491–506.

Varenne, Franck. *From Models to Simulations.* New York: Routledge, 2019.

Veit, Ulrich. "Archäologiegeschichte als Wissenschaftsgeschichte: Über Formen und Funktionen historischer Selbstvergewisserung in der Prähistorischen Archäologie." *Ethnographisch-Archäologische Zeitschrift* 52 (2011): 34–58.

Vettel, Eric J. *Biotech: The Countercultural Origins of an Industry.* Philadelphia: University of Pennsylvania Press, 2006.

Vydra, Anton. "Gaston Bachelard and His Reactions to Phenomenology." *Continental Philosophy Review* 47 (2014): 45–58.

Wacquant, Loïc, and Aksu Akçaoglu. "Pratique et pouvoir symbolique chez Bourdieu vu de Berkeley." *Revue de l'Institut de Sociologie (ULB)* 86 (2016): 35–60.

Wainwright, Milton. "Early History of Microbiology." *Advances in Applied Microbiology* 52 (2003): 333–55.

Warburg, Otto. "Über die katalytischen Wirkungen der lebendigen Substanz." In *Über die katalytischen Wirkungen der lebendigen Substanz: Arbeiten aus dem Kaiser Wilhelm-Institut für Biologie Berlin-Dahlem,* 1–13. Berlin: Julius Springer, 1928.

Watson, James D. *Molecular Biology of the Gene.* New York and Amsterdam: W. A. Benjamin, 1965.

Watson, James D. "The Synthesis of Proteins upon Ribosomes." *Bulletin de la Société de Chimie Biologique* 46 (1964): 1399–1425.

Wendler, Reinhard. *Das Modell zwischen Kunst und Wissenschaft.* Munich: Fink, 2013.

Wendler, Reinhard. "Visuelles Denken in James Watsons 'The Molecular Biology of the Gene.'" In *Long Lost Friends: Zu den Wechselbeziehungen zwischen Design-, Medien- und Wissenschaftsforschung,* ed. Christoph Windgätter and Claudia Mareis, 23–37. Zurich: Diaphanes, 2013.

Werner, Petra. *Otto Warburg: Von der Zellphysiologie zur Krebsforschung.* Berlin: Verlag Neues Leben, 1988.

White, Hayden. *The Content of the Form: Narrative Discourse and Historical Representation.* Baltimore, MD and London: Johns Hopkins University Press, 1987.

Whiting, Phineas W. "Rearing Meal Moths and Parasitic Wasps for Experimental Purposes." *The Journal of Heredity* 12 (1921): 255–61.

Willstätter, Richard, and Margarete Rohdewald. "Über enzymatische Systeme der Zucker-Umwandlung im Muskel." *Enzymologia* 8 (1940): 1–63.

Wind, Edgar. *Experiment and Metaphysics: Towards a Resolution of the Cosmological Antinomies.* Translated by Cyril Edwards. Oxford: European Humanities Research Centre of the

University of Oxford, 2001. Originally published as *Das Experiment und die Metaphysik* (Tübingen: Mohr, 1934).

Wind, Edgar. "Some Points of Contact between History and the Natural Sciences." In *Philosophy and History: Essays Presented to Ernst Cassirer*, ed. Raymond Klibansky and Herbert James Paton, 255–64. Oxford: Clarendon Press, 1936.

Winther, Rasmus Grønfeldt. *When Maps Become the World*. Chicago: University of Chicago Press, 2020.

Wirth, Uwe. "Zitieren Pfropfen Exzerpieren." In *Kreativität des Findens—Figurationen des Zitats*, ed. Martin Roussel in cooperation with Christina Borkenhagen, 79–98. Munich: Fink, 2012.

Wirth, Uwe, and Veronika Sellier, eds. *Impfen, Pfropfen, Transplantieren*. Berlin: Kadmos, 2011.

Wise, Norton. "Making Visible." *Isis* 97 (2006): 75–82.

Witkowski, Jan A. "Alexis Carrel and the Mysticism of Tissue Culture." *Medical History* 23 (1979): 279–96.

Wittmann, Barbara, ed. *Spuren erzeugen: Zeichnen und Schreiben als Verfahren der Selbstaufzeichnung*. Zurich and Berlin: Diaphanes, 2009.

Wittmann, Barbara, ed. *Werkzeuge des Entwerfens*. Zurich and Berlin: Diaphanes, 2018.

Woodward, James F. "Data and Phenomena: A Restatement and Defense." *Synthese* 182 (2011): 165–79.

Worliczek, Hanna Lucia. *Wege zu einer molekularisierten Bildgebung: Eine Geschichte der Immunfluoreszenzmikroskopie als visuelles Erkenntnisinstrument der modernen Zellbiologie (1959–1980)*. PhD diss., University of Vienna, 2020.

Yaneva, Albena. "Scaling Up and Down: Extraction Trials in Architectural Design." *Social Studies of Science* 35 (2005): 867–94.

Yang, Lei, Valentina Perrera, Eleftheria Saplaoura, Federico Apelt, Mathieu Bahin, Amira Kramdi, Justyna Olas, Bernd Müller-Roeber, Ewelina Sokolowska, Wenna Zhang, Runsheng Li, Nicolas Pitzalis, Manfred Heinlein, Shoudong Zhang, Auguste Genovesio, Vincent Colot, and Friedrich Kragler. "m^5C Methylation Guides Systemic Transport of Messenger RNA Over Graft Junctions in Plants." *Current Biology* 29 (2019): 2465–76.

Yi, Doogab. *The Recombinant University: Genetic Engineering and the Emergence of Stanford Biotechnology*. Chicago: University of Chicago Press, 2015.

Yonath, Ada, William S. Bennett, Shulamith Weinstein, and Heinz Günter Wittmann. "Crystallography and Image Recontructions of Ribosomes." In *The Ribosome: Structure, Function, and Evolution*, ed. Walter E. Hill, Albert Dahlberg, Roger A. Garrett, Peter B. Moore, David Schlessinger, and Jonathan R. Warner, 134–47. Washington, DC: ASM Press, 1990.

Zamecnik, Paul. "Historical and Current Aspects of the Problem of Protein Synthesis." In *The Harvey Lectures 1958–59*, 256–81. New York and London: Academic Press, 1960.

Zamecnik, Paul. "Historical Aspects of Protein Synthesis." *Annals of the New York Academy of Sciences* 325 (1979): 269–301.

Zilsel, Edgar *Die Geniereligion. Ein kritischer Versuch über das moderne Persönlichkeitsideal, mit einer historischen Begründung*. Edited by Johann Dvorak. Frankfurt am Main: Suhrkamp, 1990. First published 1918.

Zilsel, Edgar. "Geschichte und Biologie, Überlieferung und Vererbung." In *Wissenschaft und Weltauffassung. Aufsätze 1929–1933*, ed. Gerald Mozetic, 101–44. Vienna: Böhlau, 1998. First published in *Archiv für Sozialwissenschaft und Sozialpolitik* 65 (1931).

Index of Names

Alechinsky, Pierre, 188n23

Bachelard, Gaston, 5, 13–14, 71–72, 129–31, 134–35, 137–40, 149, 156–63, 174–76
Badiou, Alain, 32
Bastide, Roger, 156
Baudrillard, Jean, 45
Beijerinck, Martinus Willem, 112
Bensley, Robert, 123
Benveniste, Emile, 137, 144
Benzer, Seymour, 189n4
Bernard, Claude, 82–83, 112, 118, 142–43, 189n4
Berz, Peter, 82
Bloch, Marc, 205n29
Blumenberg, Hans, 6, 84, 132, 141–42
Bourdieu, Pierre, 71, 128, 149
Brachet, Jean, 29, 123
Brenner, Sydney, 104
Brunschvicg, Léon, 13
Buchner, Eduard, 117, 120

Canguilhem, Georges, 39, 145–46, 150, 161, 169
Carrel, Alexis, 121
Caspari, Ernst, 111, 121
Cassirer, Ernst, 4, 6, 101, 156, 158, 162
Cavaillès, Jean, 13, 142, 162
Chargaff, Erwin, 23
Claude, Albert, 29, 123
Correns, Carl, 106, 146
Coulson, Alan, 16
Crick, Francis, 104
Crookes, William, 15
Cuvier, Georges, 165, 171

Dalgliesh, Charles, 29
Darwin, Charles, 110, 171

Delbrück, Max, 59
Derrida, Jacques, 12–13, 70, 100, 136
Desanti, Jean-Toussaint, 6, 202n21
Descartes, René, 158, 162
Didi-Huberman, Georges, 12
Dohrn, Carl August, 110
Du Bois-Reymond, Emil, 50, 89

Enriques, Federigo, 161–62

Febvre, Lucien, 205n29
Fernbach, Auguste, 119
Feyerabend, Paul, 161
Fleck, Ludwik, 25, 103–4, 172
Flocon, Albert, 174–76, 202n12
Foucault, Michel, 5, 94, 100, 150
Frege, Gottlob, 39
Friedmann, Herbert, 117–19
Fritsch, Edward, 85

Galilei, Galileo, 132, 198n3
Galison, Peter, 166
Geitel, Hans, 15
Ginzburg, Carlo, 146–47
Goethe, Johann Wolfgang von, 47, 166–67
Gould, George, 116
Grene, Marjorie, 139
Grmek, Mirko, 83
Gros, François, 51

Hacking, Ian, 141
Harrison, Ross, 120–21
Hase, Albrecht, 110–11
Heider, Fritz, 73
Heisenberg, Werner, 113
Helmholtz, Hermann von, 64

Printed and bound by CPI Group (UK) Ltd, Croydon, CR0 4YY

08/05/2023

03216573-0001